Preface

Trace metals occur as natural constituents of the earth's crust, and are ever present constituents of soils, natural waters and living matter. The biological significance of this disparate assemblage of elements has gradually been uncovered during the twentieth century; the resultant picture is one of ever-increasing complexity. Several of these elements have been demonstrated to be essential to the functions of living organisms, others appear to only interact with living matter in a toxic manner, whilst an ever-decreasing number do not fall conveniently into either category.

When the interactions between trace metals and plants are considered, one must take full account of the known chemical properties of each element. Consideration must be given to differences in chemical reactivity, solubility and to interactions with other inorganic and organic molecules. A clear understanding of the basic chemical properties of an element of interest is an essential pre-requisite to any subsequent consideration of its biological significance. Due consideration to basic chemical considerations is a theme which runs through the collection of chapters in both volumes.

Perhaps the single most important stimulus to the rapid expansion of research in this field has been the great strides made in analytical techniques, particularly during the past decade. In many respects these advances have proceeded more rapidly than our ability to place the resultant data in its proper perspective; each increase in analytical sensitivity or reduction of detection limits has highlighted the inherent problems of sample contamination. In addition, the present dependence of most readily-available techniques on liquid samples results in the production of complex analytical matrices which present analytical problems for several elements. The need to overcome this problem is one area in which future developments will be keenly awaited. A further, more disturbing aspect of basic analytical procedures is the lack of rigorous verification of individual methodology; many inter-laboratory comparisons often reveal significant discrepancies in analytical precision on the same prepared samples. As analytical data is the core of the majority of research in this field, all workers should pay more than lip service to the validity of their analytical techniques.

Having summarized the inherent problems in this field of study, one must now consider the best way in which the ever-expanding literature on this topic can be treated, to produce a meaningful compendium of information. Any editor is faced with a difficult dichotomy in the organization of the subject matter. It is now conclusively established that the impact of trace metals on plants, in real world situations, is never the result of the effect of a single element. Thus, it could be cogently argued that treatment of individual plant/element responses, relating, in the main, to laboratory or glasshouse studies, is not a particularly valid approach to the topic. Nevertheless, a great deal of valuable information can still be gathered from this approach; the study of elemental interactions is still in a formative stage, with some being well known and clearly understood, whereas others, only recently identified, are still under careful investigation. At present, therefore, the adoption of the 'element in isolation' approach to be found in Volume 1, has been taken as a matter of necessity. Hopefully, in the next decade, this will not be the case, and workers will lean much more towards an interactive approach; but for the present, this compromise situation is the only feasible approach.

The individual elements, explored in Volume 1, illustrate the wide diversity of plant response to trace metals. As technology continues to concentrate previously dispersed and rare elements, removing these from their original matrices and altering their chemical form, great care will have to be taken in interpreting the potential consequences of these steps to living organisms. The information presented here illustrates many interesting facets of the basic interaction. Elements such as copper and zinc, long known as essential to life, are now being joined in this category by nickel. The alarming mobility of cadmium in soils and crops is highlighted, but the other toxic element, lead, is demonstrated to have a much lesser potential impact on crop production. Many elements will need to be exhaustively appraised as new directions in manufacturing industry and energy generation provide the potential to liberate more exotic and little-considered elements into the biosphere. These should be considered now, before problems arise.

One certain aspect of this area is that the consequences of metal

£21-00

EFFECT OF HEAVY METAL POLLUTION ON PLANTS

Volume 2

Metals in the Environment

POLLUTION MONITORING SERIES

EFFECT OF HEAVY METAL POLLUTION ON PLANTS

Volume 2

Metals in the Environment

Edited by

N. W. LEPP

Department of Biology, Liverpool Polytechnic, Liverpool, UK

APPLIED SCIENCE PUBLISHERS
LONDON and NEW JERSEY

APPLIED SCIENCE PUBLISHERS LTD
Ripple Road, Barking, Essex, England
APPLIED SCIENCE PUBLISHERS, INC.
Englewood, New Jersey 07631, USA

British Library Cataloguing in Publication Data

Effect of heavy metal pollution on plants.—
(Pollution monitoring series)
Vol. 2: Metals in the environment
1. Plants, Effects of heavy metals on
I. Lepp, N. W. II. Series
581.5′222 QK753.H4

ISBN 0-85334-923-1

WITH 67 TABLES AND 24 ILLUSTRATIONS

Printed in Great Britain by Galliard (Printers) Ltd, Great Yarmouth contamination will be with us for a considerable period of time. Thus, the need for more critical identification of potential future problems continues to exist and such studies should place particular emphasis on agriculture and food production. Lessons learnt in the past must be rigorously applied to future research considerations, and areas where the magnitude of potential problems has been overstated must be carefully re-appraised.

The second volume presents a more integrated approach to plant/trace metal interactions, but the majority of evidence presented here requires the sound base of information from individual studies of the 'plant–element' type described in Volume 1.

It is essential that we understand the sources of trace metal additions to our environment, the reaction these have with soils, their biogeochemical cycles, and their consequences for natural and agricultural systems. All these topics have received detailed coverage, and the synthesis of ideas presented will be of considerable relevance to those concerned with the monitoring, regulation and amelioration of trace metal contamination. Practical aspects of reclamation are also emphasized, with full documentation of successful case histories.

Finally, two topics deserved special attention, the tolerance of plants to certain trace metals, and the interaction between trace metals and lower terrestrial plants. Metal tolerance is a phenomenon often noted, but the understanding of the physiological and biochemical aspects of this fascinating aspect of plant metabolism have not progressed apace. Recent advances in our understanding of this biological puzzle are carefully detailed. Lower plants have an interesting interrelationship with trace metals. The infinite capacity of some bryophytes and lichens to sorb trace metals has led to their use as biological monitors for atmospheric metal burdens; the unique lichen symbiosis presents a system in which basic effects of metals on biological systems can be monitored and probed.

In the compilation of a work of this nature, it is inevitable, and indeed desirable, that areas of overlap will occur. The juxtaposition of several points of view in relation to a topic is a valuable part of any critical appraisal; thus no attempt has been made to reconcile any differences of opinion. The authors have been selected for their specialized knowledge of particular fields, and all have contributed to produce an integrated synthesis of the current status of plant/trace metal interactions. I would like to record my pleasure in editing these volumes, and to praise the consistently high standard and informative nature of each contribution. I would also like to thank the authors for their time, effort and ready co-operation in the assembly of this work. It is they, not I, who have translated an idea into reality. Finally, I would like to thank my colleagues at Liverpool Polytechnic for their encouragement and advice, and also the publishers for their full and friendly assistance in all stages of the production of these volumes.

NICHOLAS W. LEPP

Contents

List of Contributors

M. A. S. BURTON

Atmospheric Chemistry Division, Air Quality and Inter-Environmental Research Branch, Atmospheric Environment Service, Environment Canada, 4905 *Dufferin Street, Downsview, Ontario, Canada M*3*H* 5*T*4

P. J. COUGHTREY

*Department of Botany, University of Bristol, Woodland Road, Bristol BS*8 1*UG, UK*

B. FREEDMAN

*Department of Biology and Institute for Resource and Environmental Studies, Dalhousie University, Halifax, Nova Scotia, Canada B*3*H* 4*J*1

M. K. HUGHES

*Department of Biology, Liverpool Polytechnic, Byrom Street, Liverpool L*3 3*AF, UK*

T. C. HUTCHINSON

*Department of Botany and Institute for Environmental Studies, University of Toronto, Toronto, Ontario, Canada M*5*S* 1*A*1

M. S. JOHNSON

Department of Botany, University of Liverpool, P.O. Box 147, *Liverpool L*69 3*BX, UK*

M. H. MARTIN

Department of Botany, University of Bristol, Woodland Road, Bristol BS8 1UG, UK

K. J. PUCKETT

Atmospheric Chemistry Division, Air Quality and Inter-Environmental Research Branch, Atmospheric Environment Service, Environment Canada, 4905 Dufferin Street, Downsview, Ontario, Canada M3H 5T4

I. THORNTON

Applied Geochemistry Research Group, Department of Geology, Imperial College of Science and Technology, Royal School of Mines, Prince Consort Road, London SW7 2BP, UK

D. A. THURMAN

Department of Botany, University of Liverpool, P.O. Box 147, Liverpool L69 3BX, UK

J. WEBBER

Formerly Ministry of Agriculture, Fisheries and Food, Government Buildings, Lawnswood, Leeds LS16 5PY, UK

A. WILLIAMSON

Department of Botany, University of Liverpool, P.O. Box 147, Liverpool L69 3BX, UK

CHAPTER 1

Geochemical Aspects of the Distribution and Forms of Heavy Metals in Soils

I. THORNTON

Applied Geochemistry Research Group, Department of Geology, Imperial College, London, UK

1. INTRODUCTION

The total concentrations of trace metals and metalloids in soils, their chemical forms, mobility and availability to the food chain, provide the basis for a range of problems in crop, animal and human health. Some 15 elements present in rocks and soils, normally in very small amounts, are essential for plant and/or animal nutrition. Boron, copper, iron, manganese, molybdenum, silicon, vanadium and zinc are required by plants; copper, cobalt, iodine, iron, manganese, molybdenum, selenium and zinc by animals. The roles of arsenic, fluorine, nickel, silicon, tin and vanadium have also been established in recent years in animal nutrition. In large concentrations, many of the trace elements/metals may be toxic to plants and/or animals, or may affect the quality of foodstuffs for human consumption. These potentially toxic elements include arsenic, boron, cadmium, copper, fluorine, lead, mercury, molybdenum, nickel, selenium and zinc.

The natural concentration ranges of most trace metals in soils are wide. The main sources are the parent materials from which the soils are derived. These are usually weathered bedrock or overburden transported by wind, water or glaciation, which may be of local or exotic origin. However, man-effected inputs may add to, and at times exceed, those from natural geological sources. The main sources of metal contaminants in soils are from metalliferous mining and smelting activities, other industrial emissions and effluents, urban development, vehicle emissions, dumped waste materials, contaminated dusts and rainfall, sewage sludge, pig slurry, composted town refuse, fertilizers, soil ameliorants and pesticides.

Both deficiencies and excesses of trace metals may give rise to nutritional or toxicological problems in plants and animals, resulting sometimes in crop failure or death of farm animals. Less severe imbalance may result in a lowering of crop or animal production. Sub-clinical effects, where visual symptoms are absent, may be present but are frequently not recognized. These latter effects are thought to have considerable economic implications, as the land areas involved are likely to be large.

The soil is the primary supplier of trace metals to the soil–plant–animal system and the soil–foodstuff/water–human system. In these systems, the metals and metalloids do not of course occur in isolation, and a number of synergistic and antagonistic interactions are recognized at both deficiency and excess concentrations. These interactions sometimes involve major elements as well as trace metals, as illustrated by the effect of calcium ions on the specific adsorption of cadmium onto root surfaces and the copper–molybdenum–sulphur interrelationship in ruminant nutrition. In this chapter, specific reference will be made to the elements copper, selenium, molybdenum, arsenic, zinc, lead and cadmium which are either of economic significance in the above systems or of concern as environmental pollutants at this present time. These elements will, however, be treated as part of their association with other elements, of both geochemical and industrial origin, and emphasis will be placed on interaction at the soil–plant interface.

Information on the sources of trace metals in soils and plants and on the regional distribution of trace metal problems in agricultural crops and livestock is presented in greater detail in recent reviews by Thornton and Webb (1980*a*,*b*).

2. SOURCES OF TRACE METALS IN SOILS

Sources of metals in soils may be from natural geological materials or may arise from man's activities. In terms of environmental pollution, it is essential to have reliable data on the naturally occurring amounts of metals present in order to assess the significance of contributions from man and industry. The normal abundance of an element in earth material is commonly referred to by the geochemist as *background*, and for any particular element this value, or range of values, is likely to vary according to the nature of the material (Hawkes and Webb, 1962). The establishment

TABLE 1.1

RANGE AND MEAN CONCENTRATIONS OF SOME METALS AND METALLOIDS IN IGNEOUS AND SEDIMENTARY ROCKS (ppm)

Element	Basaltic igneous	Granitic igneous	Shales and clays	Black shales	Limestones	Sandstones
As	0·2–10	0·2–13·8	—	—	0·1–8·1	0·6–9·7
	(2·0)[a]	(2·0)	(10)	—	(1·7)	(2·0)
Cd	0·006–0·6	0·003–0·18	0–11	<0·3–8·4	—	—
	(0·2)	(0·15)	(1·4)	(1·0)	(0·05)	(0·05)
Cr	40–600	2–90	30–590	26–1 000	—	—
	(220)	(20)	(120)	(100)	(10)	(35)
Co	24–90	1–15	5–25	7–100	—	—
	(50)	(5)	(20)	(10)	(0·1)	(0·3)
Cu	30–160	4–30	18–120	20–200	—	—
	(90)	(15)	(50)	(70)	(4)	(2)
Hg	0·002–0·5	0·005–0·4	0·005–0·51	0·03–2·8	0·01–0·22	0·001–0·3
	(0·05)	(0·06)	(0·09)	(2·5)	(0·04)	(0·05)
Pb	2–18	6–30	16–50	7–150	—	<1–31
	(6)	(18)	(20)	(30)	(9)	(12)
Mo	0·9–7	1–6	—	1–300	—	—
	(1·5)	(1·4)	(2·5)	(10)	(0·4)	(0·2)
Ni	45–410	2–20	20–250	10–500	—	—
	(140)	(8)	(68)	(50)	(20)	(2)
Se	—	—	—	—	—	—
	(0·05)	(0·05)	(0·6)	—	(0·08)	(0·05)
Zn	48–240	5–140	18–180	34–1 500	—	2–41
	(110)	(40)	(90)	(100)	(20)	(16)

[a] Numbers in parentheses are mean concentrations.
Adapted from table compiled by M. Fleischer and H. L. Cannon (Cannon *et al.*, 1978).

of background values for metals in soils and plants has for some time been recognized as part of the geochemist's role in pollution studies (Cannon and Anderson, 1971).

2.1. Weathering of Unmineralized Bedrock and Other Parent Materials

The earth's crust is made up of 95 % igneous rocks and 5 % sedimentary rocks; of the latter about 80 % are shales, 15 % sandstones and 5 % limestones (Mitchell, 1964). However, sediments are more frequent at the surface as they tend to overlie the igneous rocks from which they were derived. The abundance of some metals and metalloids in igneous and sedimentary rocks is shown in Table 1.1.

TABLE 1.2

THE CADMIUM CONTENT (ppm) OF SOME BLACK SHALES IN ENGLAND AND WALES. (AFTER HOLMES, 1975)

Formation	Locality	Age	Range	Mean[a]
Lower Worston shale group	Bowland Forest, Lancs	B1–2	<1–32	4·4 (46)
Lower Bowland shale group	Bowland Forest	P1	1–105	16·2 (35)
Lower Bowland shale group	Bowland Forest	P2	1–158	16·5 (20)
Upper Bowland shale group	Bowland Forest	E1	1–219	16·6 (59)
Edale shales	North Derbyshire	E2	1–39	5·2 (48)
Edale shales	North Derbyshire	H1	1–50	6·0 (14)
Edale shales	North Derbyshire	H2	<1–91	14·8 (25)
Edale shales	North Derbyshire	R1	1–32	6·0 (11)
Dove shales	South-west Derbyshire	E2	<1–25	6·5 (45)
Mixon limestone and shales	North Staffordshire	P1–2	<1–65	12·8 (31)
Onecote sandstone and shales	North Staffordshire	P1–2	1–39	9·3 (8)
Onecote sandstone and shales	North Staffordshire	E2	1–2	1·4 (12)
Cracklington formation	Devon/Cornwall	H1	<1–5	1·3 (39)
Cracklington formation	Devon/Cornwall	R1	<1–4	1·7 (55)
Cracklington formation	Devon/Cornwall	R2	<1–3	1·5 (29)
Coal measures	Glamorgan	d5	<1–5	1·0 (9)
Coal measures	Chesterfield	d5	<1–3	1·5 (15)

[a] Number of samples in parentheses.

The degree to which trace elements become available upon the weathering of igneous rocks, depends on the type of minerals in which they are present and the susceptibility of these minerals to weathering. The more biologically important trace metals, including copper, cobalt, manganese and zinc, occur mainly in the more easily weathered constituents of igneous rocks such as augite, hornblende and olivine (Mitchell, 1974). Of the sedimentary rocks, sandstones are composed of minerals that weather with difficulty and usually contain only small amounts of trace metals. Shales, on the other hand, may be of inorganic or organic origin, and usually contain larger amounts of trace metals (Mitchell, 1964). It is seen that black shales in particular are enriched in copper, lead, zinc, molybdenum and mercury (Table 1.1). Detailed studies on cadmium in British black shales showed a wide range of concentrations in those examined ranging up to 219 ppm (Table 1.2; Holmes, 1975).

Soils developed from these parent materials tend to reflect their chemical composition, though pedogenetic processes may modify this relationship. Soils derived from the weathering of coarse-grained materials such as sands and sandstones and from acid igneous rocks such as rhyolites and granites, tend to contain smaller amounts of nutritionally essential metals, including copper, cobalt and zinc, than do those derived from fine-grained sedimentary rocks such as clays and shales, and from basic igneous rocks. For example, the average copper content of British surface soils has been recorded as 20 ppm (Swaine and Mitchell, 1960). However, the total copper contents of a large number of surface soils from England and Wales (Table 1.3) reflect a wide range, from 2 ppm in soils derived from Pleistocene sands in East Anglia, to 2000 ppm in both alluvial and upland soils in mineralized areas of south-west England. Detailed studies on soils

TABLE 1.3

THE TOTAL COPPER CONTENTS OF SOILS DEVELOPED FROM A VARIETY OF PARENT MATERIALS IN ENGLAND AND WALES

Geological formation	Cu (ppm)
Recent	
Marine silt	9–50[e]
(East Anglia)	**22** (29)
Pleistocene	
Breckland	2–20
(chalk–sand drift)	**8** (35)
(East Anglia)	
Tertiary	
Bagshot Beds	4–37[a]
(Dorset)	**13** (22)
Cretaceous	
Chalk	7–28[e]
(Wiltshire)	**15** (128)
(Sussex and Hants)	12–17[e]
	14 (3)
(Berks and Oxon)	10–17[e]
	13 (7)
Upper Greensand	
(Sussex and Hants)	10–20[e]
	14 (8)
(Berks and Oxon)	7–14[e]
	11 (3)
(Wilts)	4–18[e]
	10 (28)
Gault	
(Sussex and Hants)	10–17[e]
	13 (9)
(Berks and Oxon)	14–25[e]
	19 (6)
Lower Greensand	
Hythe Beds	3–13[e]
(Sussex and Hants)	**8** (28)
Sandgate Beds	4–14[e]
(Sussex and Hants)	**9** (26)
Folkestone Beds	2–17[e]
(Sussex and Hants)	**8** (28)
Weald Clay	
(Sussex and Hants)	10–25[e]
Jurassic	
Kimmeridge Clay	9–27[e]
(Berks and Oxon)	**16** (21)
(south-west England)	11–40[c]
	18 (30)
Corrallian Sandstone	5–19[e]
(Berks and Oxon)	**11** (30)
Corrallian Limestone	10–25[e]
(Berks and Oxon)	**16** (15)
(south-west England)	11–26[e]
	17 (11)
Oxford Clay	
(Berks and Oxon)	10–26[e]
	20 (12)
(south-west England)	10–25
	18 (10)
Cornbrash	14–20[c]
(south-west England)	**17** (4)
Lower Lias	30–40[d]
(Gloucestershire)	**35** (6)

(continued overleaf)

TABLE 1.3—*contd.*

Geological formation	Cu (ppm)	Geological formation	Cu (ppm)
(Somerset)	6–60[a] **32** (30)	Downton Marls (Herefordshire)	16–60[d] **30** (56)
Permo-Triassic			
Permian Sandstone (Cumberland)	5–40[d] **15** (18)	Raglan Marl (Monmouth)	20–40[d] **28** (19)
Triassic Sandstone (Cumberland)	4–16[d] **10** (17)	St Maughans Group (Monmouth)	10–40[d] **21** (4)
Keuper and Bunter Sandstone (Lower Severn Valley)	10–40[d] **20** (13)	Brownstone (Monmouth)	5–30[d] **18** (37)
Keuper Marl (Lower Severn Valley)	10–300[d] **48** (80)	Tintern Sandstone (Monmouth)	5–20[d] **13** (24)
Carboniferous shales (Staffordshire)	**40** (68)[f]	Ditton Sandstone (Hay-on-Wye)	20–50[d] **32** (56)
Devonian		*Silurian* (Denbigh Upland)	9–90[b] **22** (99)
Devonian Slates (South Devon)	14–42[b] **26** (44)	(Denbigh Moorland)	6–32[b] **15** (45)
Devonian Sandstones (North Devon)	2–31[b] **14** (37)	(Herefordshire)	20–50[d] **31** (15)
Old Red Sandstone	13–40[d] **27** (60)		

Soils contaminated by mining and smelting in the Tamar Valley area of south-west England:

Upland Soils	29–2000[g] ppm Cu **314** (28)
Alluvial Soils	35–2000[g] ppm Cu **620** (12)

[a] Thornton (1968)
[b] Keeley (1972)
[c] Thomson (1971)
[d] Wood (1975)
[e] Jordan (1975)
[f] Fletcher (1968)
[g] Colbourn *et al.* (1975)
Number of samples in parentheses.
Bold type indicates mean value.

developed from individual beds within selected sandstone formations have shown that the total copper content of the soil is generally lowest on the coarser grained parent materials and that concentrations increase with decreasing grain size (Table 1.4; Wood, 1975).

On the other hand, potentially toxic amounts of trace metals in soils may be derived from naturally occurring metal-rich source rocks (Table 1.5). Nickel-rich soils derived from ultra-basic rocks containing ferro–magnesium minerals in parts of Scotland may lead, under poor drainage

TABLE 1.4

AVERAGE COPPER CONTENT AND TEXTURE OF SOILS DERIVED FROM INDIVIDUAL BEDS OF THE OLD RED SANDSTONE FORMATION (UPPER DEVONIAN) IN SOUTH WALES (WOOD, 1975)

Bed	Soil texture			Total Cu
	Coarse sand	Fine sand	Silt and clay	(ppm)
Raglan Marl	14	50	36	28
St Maughans Group	13	60	27	21
Brownstones	66	14	20	18
Tintern Sandstones	51	25	24	13

TABLE 1.5

TRACE ELEMENTS IN SOILS DERIVED FROM NORMAL AND GEOCHEMICALLY ANOMALOUS PARENT MATERIALS

Normal range in soil (ppm)		Range in metal-rich soils (ppm)	Sources	Possible effects
As	< 5–40	up to 2 500	Mineralization	Toxicity in plants and livestock; excess in food crops
		up to 250	Metamorphosed rocks around Dartmoor	
Cd	< 1–2	up to 800	Mineralization	Excess in food crops
		up to 20	Carboniferous black shale	
Cu	2–60	up to 2 000	Mineralization	Toxicity in cereal crops
Mo	< 1–5	10–100	Marine black shales of varying age	Molybdenosis or molybdenum-induced hypocuprosis in cattle
Ni	2–100	up to 8 000	Ultra-basic rocks in Scotland	Toxicity in cereal and other crops
Pb	10–150	1 % or more	Mineralization	Toxicity in livestock; excess in foodstuffs
Se	< 1–2	up to 7	Marine black shales in England and Wales	No effect
		up to 500	Namurian shales in Ireland	Chronic selenosis in horses and cattle
Zn	25–200	1 % or more	Mineralization	Toxicity in cereal crops

conditions, to nickel toxicity in cereal and other crops (Mitchell, 1974). Of particular significance to agriculture in Britain, is the observation that excess molybdenum in soils and pastures can give rise to molybdenosis or molybdenum-induced copper deficiency in cattle. Molybdenum toxicity was originally described on calcareous soils containing 20 ppm Mo or more derived from the Lower Lias formation in Somerset (Ferguson *et al.*, 1943; Lewis, 1943; Le Riche, 1959). At the present time it is recognized that soils containing 5 ppm Mo or more may support herbage containing 2 ppm Mo or more; this may result in loss of production and growth retardation in cattle due to molybdenum-induced copper deficiency. In England and Wales, soils derived from marine black shales of Cambrian, Ordovician, Silurian, Jurassic, Carboniferous, Cretaceous and Recent age usually contain from 1 to 100 ppm Mo compared with the majority of soils developed on other parent materials with less than 2 ppm Mo (Thomson *et al.*, 1972; Thornton and Webb, 1976).

The influence of parent materials on the total content and form of trace metals in soils is modified to varying degrees by soil-forming processes, which may lead to the mobilization and redistribution of elements both within the soil profile and between neighbouring soils (Swaine and Mitchell, 1960; Mitchell, 1964). In the UK and similar temperate areas, most soils are relatively young and the parent material remains the dominant factor. Under tropical climates and on more mature land surfaces, e.g. Australia, weathering processes have been more vigorous or of much greater duration, and relationships between the chemical composition of the original parent materials and the soil may be completely changed by the mobilization and secondary distribution of chemical elements and the formation of secondary minerals.

The processes of gleying, leaching, surface organic matter accumulation and podzolization, together with properties such as reaction (pH) and redox potential (E_h), may affect the distribution, the form and the mobility of trace elements in the soil. The mobilization of trace metals by aerobically decomposing plant material has been described by Bloomfield (1969), and it has been shown that mobilization is partly in association with colloidal humidified organic matter, and partly in true solution as complexes that seem to be anionic (Bloomfield *et al.*, 1976). The distribution of trace elements in freely drained and poorly drained Scottish soils has been described by Mitchell (1971). The solubility and mobility of individual metals varies considerably. Iron, manganese, cobalt, cadmium and zinc are relatively mobile compared to lead and molybdenum and are sometimes redistributed in the course of soil formation and maturity. Redistribution of cadmium and zinc in soils derived from Carboniferous marine shale in Derbyshire is shown in Table 1.6. In contrast to lead, both metals have been leached from freely and imperfectly drained soils on hill tops and slopes and accumulated under reducing conditions in the lower horizons of very poorly drained soils at the base of slopes (Marples, 1979).

TABLE 1.6

CADMIUM, ZINC AND LEAD IN SOILS DEVELOPED FROM CARBONIFEROUS BLACK SHALE IN DERBYSHIRE (MARPLES, 1979)

Depth (cm)	Top of slope			Depth (cm)	Waterlogged valley bottom		
	Cd	Zn	Pb		Cd	Zn	Pb
	(ppm dry soil)				(ppm dry soil)		
0–10	4	248	316	0–10	10	312	204
10–20	4	228	252	40–50	14	660	414
20–50	3	240	160	70–80	36	1 000	328
50–75	3	356	44	80–90	52	2 280	400
75–100	9	536	100	—	—	—	—

Where residual soils are formed *in situ* from the underlying bedrock, the trace element content of the soil may be directly related to that of the bedrock. Where parent materials have been mixed or redistributed by alluvial transport, wind or glacial activity, the influence of the underlying rock may be either modified or completely masked.

2.2. Weathering of Mineralized Rocks and Materials Derived from Them

Soils in mineralized areas frequently contain high concentrations of one or more of the metals copper, lead, zinc, cadmium and arsenic. In Britain, amounts of lead and zinc can exceed 1 %, copper may range up to 1000 ppm or more, and cadmium to 800 ppm (Table 1.7). Such soils may be developed over, or in, the dispersion halo around ore bodies and mineral veins or from transported overburden containing mineralized materials. The factors controlling metal dispersion from these sources in residual soils, glacial

TABLE 1.7

AMOUNTS OF CADMIUM, ZINC AND LEAD IN SURFACE SOILS (μg/g, 0–15 cm) IN TWO BRITISH MINERALIZED AREAS

Area	No. of samples	Cadmium	Zinc	Lead
Derbyshire				
Mining area	13	1·1–34	94–8 000	230–48 000
Control areas	5	0·9–3·8	82–240	69–290
Shipham				
Mining area	12	29–800	2 520–62 400	720–9 600
Control areas	6	2·0–10	208–740	128–344

drift, alluvial soils and organic soils have been reviewed by Hawkes and Webb (1962), Levinson (1974) and Siegel (1974).

2.3. Contamination from Mining and Smelting

Where mineralized areas have been worked, surface soils are usually contaminated to varying degrees by the mining and smelting activities. In Britain, the most extensive areas of metal-rich soils reflect contamination from base-metal mining which commenced in Roman times or earlier, with peak output in the mid nineteenth century (Alloway and Davies, 1971; Davies, 1973; Davies and Roberts, 1975; Thornton and Webb, 1975). In an area of copper–tin–arsenic mineralization in south-west England, copper and arsenic concentrations in soils range up to 2000 ppm Cu and 2500 ppm As. The variable influence of mine waste, dust and smelter fumes, together with the effects of underlying metal-rich material, results frequently in large variations in the metal content of soils on a local scale (Colbourn *et al.*, 1975; Thoresby and Thornton, 1980). Similar extensive contamination has been shown over some 250 km^2 in the west of the Cornish peninsula (Thornton *et al.*, 1979).

Contaminated soils are found in an area of some 250 km^2 in the mineralized district of Derbyshire. These contain from several hundred to several thousand ppm of lead with maximum values exceeding 3% Pb at sites close to old workings and smelters (Thornton and Webb, 1975; Colbourn and Thornton, 1978). Levels of zinc and cadmium are also raised in part of the area. The normal range of lead in uncontaminated UK soils is from 10 to 150 ppm Pb (Table 1.5).

Present-day and recent primary smelting of ore concentrates, and secondary smelting involving the recycling of waste metals, are major sources of metal contamination, usually over limited areas in the immediate vicinity of the smelters. An example of such contamination is found as a result of stack emissions from the AMAX smelter in the New Lead Belt of Missouri where surface soil (0–2·5 cm, less than 200 μm fraction) contained on average 1663 ppm Pb, 143 ppm Zn and 45 ppm Cu at 0·4 km from the smelter compared with 15–20 ppm Pb, 10–20 ppm Pb, 10–20 ppm Zn and 1·5–4 ppm Cu at 38–48 km (Jennett *et al.*, 1977). Soils downwind of the zinc–lead smelter at Avonmouth, south-west England, contain elevated levels of lead, zinc and cadmium to a distance of around 3 km. Metal concentrations fall off rapidly with increasing distance from the smelter as illustrated by the distribution of cadmium which is found at 42 ppm in surface soil (0–15 cm) at 1·5 km downwind, and 4 ppm or less at 5 km and beyond (Fig. 1.1; Marples and Thornton, 1980). The background level of

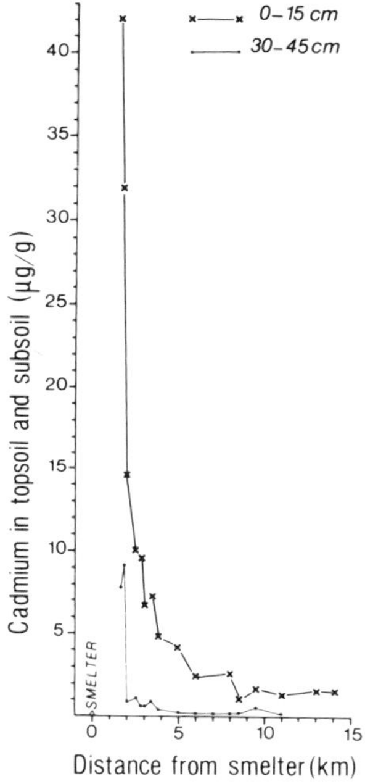

FIG. 1.1 Cadmium in surface (0–15 cm) and subsoils (30–45 cm) downwind from the zinc–lead smelter at Avonmouth. (After Marples and Thornton, 1980.)

cadmium in this case is seen to be around 1–2 ppm at the surface and less than 0·5 ppm in the subsoil of estuarine alluvium.

2.4. Other Sources of Metal Pollution

Other man-effected sources of metal contamination to soils have already been listed. A discussion of these is outside the scope of this chapter and for detailed information the reader is referred to Chapter 2 and to the following selected publications: Domestic refuse (Purves, 1977); domestic coals, coal-burning power stations and fly-ash (Swaine, 1977); sewage sludge (Le Riche, 1968; Berrow and Webber, 1972; Berggren and Oden, 1972; Blakeslee, 1973; Andersson and Nilsson, 1972, 1976; Page, 1974; Boswell, 1975; Dowdy and Larson, 1975; Furr *et al.*, 1976; Berrow and Burridge, 1980); pig slurry (Unwin, 1980); composted town refuse (Purves and Mackenzie, 1973; Gray and Biddlestone, 1980); cadmium in rock phosphate and superphosphate (Williams and David, 1973; Stenstrom and Walter, 1974; Fulkerson, 1975; Commission of the European Communities, 1978); other fertilizers and liming materials (Swaine, 1962).

3. SOIL FACTORS AFFECTING THE AVAILABILITY OF TRACE METALS TO PLANTS

Trace elements may be present in soils in primary and secondary minerals, as precipitated salts, in ionic and complexed state in the soil solution, complexed by organic materials or adsorbed onto reactive surfaces of soil constituents. The stability and weathering of common primary and secondary soil minerals has been predicted thermodynamically by Rai and Lindsay (1975). The reactivity of different sinks for trace element sorption has been discussed by Jenne (1977) who proposes that those of major significance are oxides of iron and manganese, organic matter, sulphides and carbonates; and of lesser importance are phosphates, iron salts and clay-size aluminosilicate minerals.

The availability of some metals to plants is governed by their total concentration in the soil and by the forms in which they occur. Both the actual amounts and forms of trace metals may be influenced by the processes of gleying, leaching, podzolization and surface organic matter accumulation. For example, the solubility and availability of many elements, including manganese and molybdenum, increases in poorly drained soils. Uptake by plants is also related to a range of other soil factors including pH and the presence of other major and minor elements. Manganese and cobalt availability decreases, whilst that of molybdenum usually increases with rising soil pH. However, information on the importance of these interrelationships is frequently conflicting and more detailed and systematic studies are required over a wide range of soil conditions. Earthworms, living in soil which contains large amounts of lead and zinc, can accumulate these metals, which are more easily extractable in acetic acid from the putrefied worms than from the soil (Ireland, 1975). Soil microorganisms may also play a significant role in processes governing trace metal distribution and availability, though this is a scientific area that has received little attention to date.

The availability of some metals and metalloids is discussed below.

3.1. Copper

Copper deficiencies in crop plants (Caldwell, 1971) and grazing livestock (Russel and Duncan, 1956; Underwood, 1966) have been recognized in many parts of the world; the excess of the metal can result in toxicity to ruminants (Underwood, 1971).

Copper in the soil is strongly held on inorganic and organic sites and in complexes with organic matter. As such, a large proportion of the total copper content of soils is not available for uptake by plants. Deficiencies in crops may be due to an inherently low total copper content of the soil or to only a small amount being in an available form. Deficiencies may be aggravated by soil microorganisms.

Although copper held on exchange sites is not readily available to plants, cation exchange from Cu^{2+} and $CuOH^+$ can take place, and is best effected by H^+ (Mengel and Kirkby, 1978). The amount of copper in soil solution decreases with increasing pH due to stronger copper absorption (Lindsay, 1972*a*). As pH is raised by the application of lime to soil, availability of copper to plants usually, but not always, decreases. In organic soils, availability of copper depends not only on the concentration in soil solution, but also on the form in which the copper is present (Mercer and Richmond, 1970). Copper complexes of molecular weight <1000 were found to be more available to plants than those with molecular weight exceeding 5000.

EDTA-extractable copper levels were increased in intensely gleyed horizons of Scottish soils due to increased mobilization of the metal (Mitchell, 1971). In peaty soils in eastern England, copper deficiency in crops can be prevented by maintaining a naturally high water table with irrigation, and it has been suggested that copper availability is lower where peat is dried than when it is moist (Caldwell, 1971). Copper deficiency in Britain is supposedly more severe in dry sunny years than in dull moist conditions (Caldwell, 1971).

Copper deficiency in cereal crops may be aggravated by the application of nitrogenous fertilizers (Henkens, 1957; Fleming and Delaney, 1961; Davies *et al.*, 1971), and repeated use of phosphate fertilizers may also have a similar effect (Bingham, 1963). Possible interrelationships with manganese and zinc have also been suggested (Chaudhry and Loneragan, 1970; Caldwell, 1971).

Only a small fraction of the large amounts of copper present in mine-contaminated soils in south-west England was taken up into the shoots of grass, the leaves of lettuce and the grain of barley. The relationship between soil copper and concentrations of the metal in lettuce leaf is illustrated in Fig. 1.2. This may be due to low solubility of soil copper and associated low intake into plant roots, or more likely to plant mechanisms controlling uptake and translocation (Thoresby and Thornton, 1980). The forms of copper in these heavily contaminated soils are not known.

3.2. Selenium

Both excess and deficiency of selenium in the diet result in disorders in

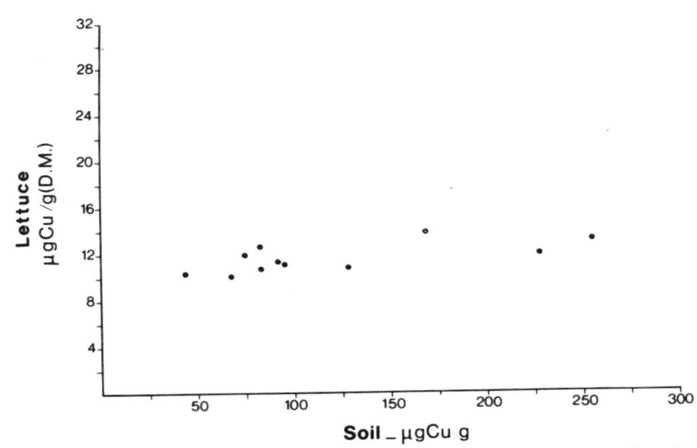

FIG. 1.2 Relationship between the copper content of lettuce leaf and the copper content of soil in the mineralized Tamar district of south-west England.

grazing livestock. In the Republic of Ireland chronic selenosis in cattle is recognized in small areas of counties Limerick, Tipperary, Meath and Dublin associated with selenium-rich parent materials derived from marine black shale rocks of Carboniferous age. Normal soils in Ireland contain 0·5–1 ppm total Se; problem soils contain from 30 to >300 ppm Se. Potentially toxic pasture ranges from 5 to 500 ppm Se in the dry matter (Fleming and Walsh, 1957; Fleming, 1962, 1968). In North America, soils associated with selenosis usually contain 2–6 ppm or more of total selenium, of which a high proportion is available to plants, probably present as selenate, selenite or organic selenium compounds. Selenium in Irish soils has a low solubility in water, and it has been shown that selenite is the predominant ion in the aqueous extract (Nye and Peterson, 1975). The relatively large uptake of selenium by pasture plants in Ireland is associated with mineral soils of limestone/shale origin with pH values in excess of 6·5, or with acid poorly drained organic soils. Peak values of up to 7 ppm total Se have been measured in mineral soils of England and Wales; pasture herbage has not been found to exceed 1 ppm Se in the dry matter (Webb *et al.*, 1971).

3.3. Molybdenum

The adverse effects of large amounts of molybdenum in soils on the health of grazing livestock has been noted earlier in this chapter. Soils developed from marine black shales of varying age may contain above normal amounts of molybdenum. Sources deriving from the activities of man include emissions and effluents from mining, aluminium alloy factories, oil refineries and coal-fired power plants and additions to soil in sewage sludge. The biogeochemistry of molybdenum in Britain has been reviewed previously (Thornton and Webb, 1976; Thornton, 1977). Several papers on molybdenum in soils and plants have been published in the Proceedings of an International Symposium on Molybdenum in the Environment held in Denver, Colorado, in 1975 (Chappell and Petersen, 1977). In particular, these proceedings contain comprehensive reviews of molybdenum in US soils and plants (Kubota, 1977; Allaway, 1977), and detailed information on the contamination of soils in Colorado by the mining of molybdenum ores and the availability of the metal to plants (Runnells *et al.*, 1977; Vlek and Lindsay, 1977).

Molybdenum is thought to occur in soils principally in a complex anionic state being: (a) precipitated by calcium; (b) in acid solution; (c) bound in organic complexes; or (d) adsorbed by anion-exchange material (Mitchell, 1964). Molybdenum may possibly be reduced by organic matter to exchangeable cationic forms (Szalay, 1964, 1969) and may be associated with clay-sized mineral particles in organo–mineral complexes (Mitchell, 1971).

The availability of soil molybdenum to plants is related to pH in inorganic soils. Uptake by grasses was found to increase over the pH range 4–7 in soils containing from 2 to 30 ppm Mo developed from black shale parent material in Wales. Uptake from soils developed from Carboniferous shales in northern England increased over the pH range 5·3–7·8 and also with organic carbon content (Thomson *et al.*, 1972). The addition of lime to soil has been found to increase molybdenum uptake by both grasses and legumes (Mitchell, 1971; Williams and Thornton, 1972). Uptake also increases with soil wetness (Mitchell, 1971). Molybdenum uptake has been inversely related to the total iron content of molybdeniferous soils in parts of England (Fletcher, 1968; Thomson *et al.*, 1972). Relatively high molybdenum uptake by grasses growing on organic soils with low pH has been recorded in Ireland (Walsh *et al.*, 1953), in Scotland (Mitchell, 1964) and in Wales (Thornton, 1968). It has been suggested that organically bound or complexed molybdenum may be available over a wide pH range.

Molybdenum uptake has been shown to be affected by soil phosphate and sulphate, though evidence is conflicting. Ammonium sulphate depressed molybdenum uptake by pasture herbage on soils developed from the Lower Lias (Lewis, 1943) and by both ryegrass and clover on a series of organic and mineral soils under greenhouse conditions (Williams and Thornton, 1972).

3.4. Zinc

Zinc in soils and in plant nutrition has been reviewed by Lindsay (1972*a*). Zinc deficiency in crops is common, though different plants vary widely in their susceptibility (Viets *et al.*, 1954). Zinc toxicities are sometimes found in grasses and crop plants growing in soils heavily contaminated with mine waste.

The total zinc content of the majority of soils greatly exceeds crop requirements, though availability is affected to a large extent by soil pH and organic matter content (Mengel and Kirkby, 1978). Deficiency is sometimes found on highly leached arid soils with total zinc concentrations of 10–30 ppm. The mean total zinc contents of soils developed over a wide textural range of parent materials in arable areas of eastern and southern England ranged from 30 to 100 ppm Zn, with smallest amounts in soils derived from sands and sandstones (Jordan *et al.*, 1975).

Zinc occurs in a number of soil minerals, the Zn^{2+} ion substituting for Fe^{2+} and Mg^{2+}; zinc may also be present as mineral salts, e.g. $ZnCO_3$, and may be adsorbed on to the surface of clay minerals and organic matter. Zinc mobility and thus availability decreases with rising soil pH and deficiencies usually occur on soils of naturally high pH, especially calcareous soils and on soils that have been heavily limed.

Soils containing large amounts of zinc, sometimes as much as 1 % Zn or more, have been found in several of the old metalliferous mining areas of England and Wales and in particular in those areas associated with lead–zinc mineralization in south-west England, mid and north Wales and in the southern and northern Pennines. Zinc in these soils usually occurs in association with lead and cadmium.

3.5. Lead

The lead content of uncontaminated soils in Britain ranges from 10 to 150 ppm Pb (Table 1.5), however, it is as a pollutant that lead poses a threat to the food chain and to man. The most severely contaminated soils in Britain are in mineralized areas where the lead ores galena and cerrusite were mined from Roman times to the beginning of the present century. Such contaminated areas are thought to exceed 4000 km^2; in Derbyshire alone lead-contaminated soils extend to some 250 km^2 of agricultural land (Colbourn and Thornton, 1978), with values usually exceeding 1000 ppm Pb in surface soils within 500 m of old surface workings, spoil heaps and smelter sites. Zinc and cadmium are also present in these soils as pollutants.

Lead added to the surface of soils in airborne contaminants does not usually leach down the profile, probably because of the absorption of Pb^{2+}

on the surfaces of clay minerals and organic colloids and to the formation of insoluble lead chelates with organic matter (Lagerwerff, 1972). Lead also accumulates in surface soils as a result of biological cycling through plants. Only a small proportion of soil lead is in a form available to plant roots and only a small amount of that taken into plant roots is translocated to the shoots (Jones and Clement, 1972). The availability of soil lead can be decreased by liming in two ways: (a) at a high soil pH lead may be precipitated as the hydroxide, phosphate or carbonate; (b) Ca^{2+} ions compete with lead for exchange sites on root and soil surfaces (Mengel and Kirkby, 1978).

3.6. Cadmium

Cadmium is present in most rocks and soils at concentrations of 1 ppm or less. It is, however, enriched in marine black shales of Carboniferous age in parts of Britain (Table 1.2). Peak concentrations in these rocks are as high as 100 ppm Cd or more and agricultural soils developed from them may contain up to 20 ppm (Holmes, 1975; Marples and Thornton, 1980). Large concentrations of cadmium (up to 30 ppm) are also found in soils in mineralized areas such as Derbyshire, where cadmium is associated with zinc in the mineral sphalerite, and again in similar or slightly higher concentrations in the vicinity of both primary and secondary smelters.

On farmland reclaimed from, and near to, old zinc mines at Shipham in Somerset, where the main ore mineral was calamine ($ZnCO_3$), soils contain from 30 to several hundred ppm Cd. Amounts of zinc in these soils are also very large, frequently exceeding 1 % (Fig. 1.3). It is of interest that even on these soils, the cadmium content of pasture herbage rarely exceeds 2 ppm (dry matter). It has been suggested that: (a) the cadmium is mostly unavailable for uptake by plants; (b) the metal does not pass into the roots of plants due to the large amounts of other metal ions, particularly zinc, present in the soil; or (c) translocation of cadmium from roots to the leaf material is restricted (Marples and Thornton, 1980). Using solution culture techniques, Jarvis *et al.* (1976) showed cadmium uptake by ryegrass to be significantly depressed by adding manganese, zinc and calcium to solutions. They concluded that Ca^{2+}, Zn^{2+} and Mn^{2+} ions competed with Cd^{2+} for exchange sites at the root surface and thus depressed cadmium uptake. Of cadmium taken into the roots of perennial ryegrass, less than 15 % was translocated to the shoots (Jarvis *et al.*, 1976).

Cadmium is relatively mobile in some soils and may be redistributed due to leaching both down the profile and between neighbouring soils in a catenary sequence as shown in Table 1.6. The form of cadmium in the soil is

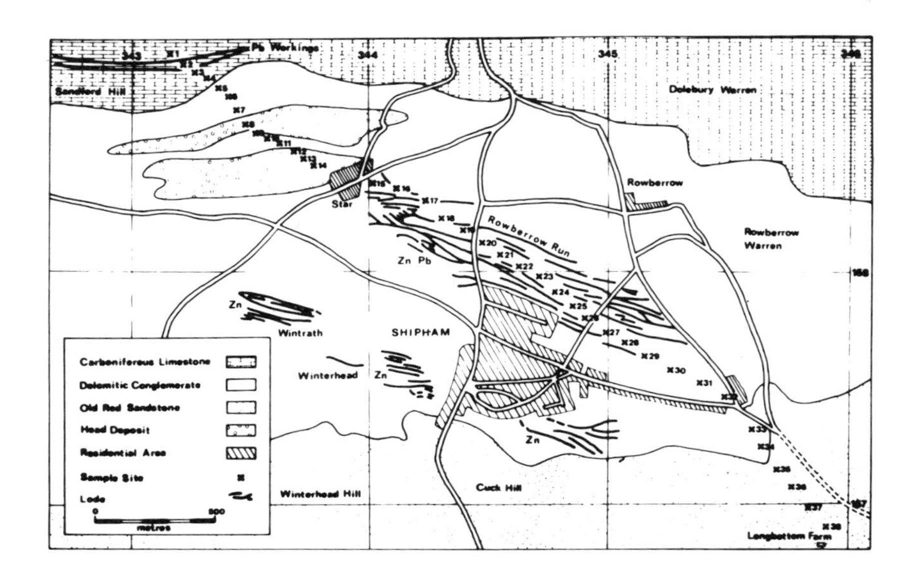

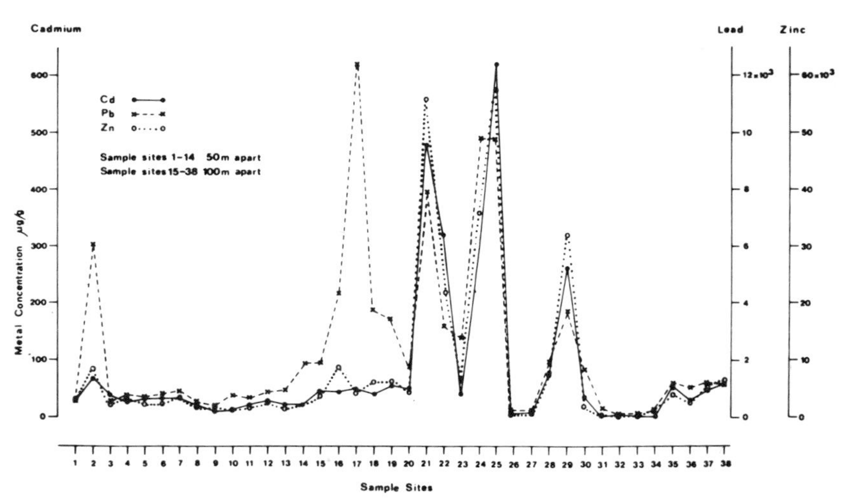

FIG. 1.3 Cadmium, zinc and lead in surface soils (0–15 cm) in the vicinity of Shipham, Somerset. (After Thornton *et al.*, 1979.)

not clearly understood, though it is likely that the metal is adsorbed onto exchange sites or is present as a metallo–organic complex. Availability and uptake by plants decreases with increasing soil pH values of 6·5 or more.

4. THE SOIL–PLANT–ANIMAL RELATIONSHIP

It has already been noted that the availability of metals, and thus uptake by plants, is related both to their total concentrations and to their forms and associations in the soil, and to a number of physical and chemical factors operating at the soil–root interface.

The influence of plant species on the soil–plant relationship may be considerable (Fleming, 1965; Archer, 1971). Different species, and indeed different cultivars, regulate metal uptake at both the soil–root and root–shoot interfaces to varying degrees. Clover tends to contain larger amounts of trace elements—particularly molybdenum—than grasses. Where soils contain excess molybdenum, the clover content of the sward should be kept low to minimize intake by grazing livestock. Accumulator plants may contain very large amounts of specific elements. *Astragalus* spp. found in some arid areas of America can contain several thousand ppm Se compared with local grasses growing on the same soil with only a few ppm. Selenium accumulator plants have not yet been found in Britain, though plants accumulating large amounts of As have been reported on mine waste in south-west England (Nye and Peterson, 1975). Metal-tolerant ecotypes of common grass species that have adapted to metal-rich soils and waste materials take up different amounts of lead and zinc to cultivars growing on uncontaminated land (Antonovics *et al.*, 1971). These tolerant plants will also be found on undisturbed mineralized soils.

In general, the trace element content of the whole plant tends to increase with increasing maturity, and seasonal influences are also important (Fleming, 1965). For example, lead was found to be present in higher concentrations in grass in winter months than in summer, on a soil of normal lead content in Scotland (Mitchell and Reith, 1966). Similar effects have more recently been found in mineralized areas of Derbyshire, where seasonal differences in the lead content of pasture were far greater than those attributable to the lead contents of the soils (Fig. 1.4; Thornton and Kinniburgh, 1978). The practical implications of this to farming are important, as amounts of lead in pasture herbage are relatively low throughout the grazing seasons for cattle. However, seasonal variations in plant composition, such as those illustrated, may not be entirely due to

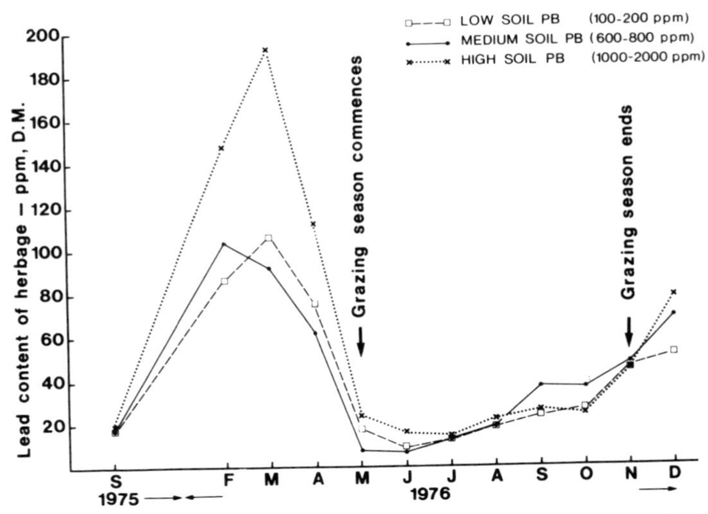

FIG. 1.4 Seasonal variations in the concentrations of lead in pasture herbage on soils of different lead contents in the mineralized district of Derbyshire, central England.

reduced uptake or redistribution of the metal within the plant; some degree of soil contamination of the plant sample is difficult to avoid, and is likely to be far greater in winter months than in the spring and summer when growth is rapid.

The effects of other major and minor elements in the soil may have a significant effect on the availability of specific trace metals. Under farming conditions, fertilizer and lime application may influence the relationship between trace metals in soils and plants. As the trace constituents of these amendments tend to be relatively small (Swaine, 1962), the effects are mainly due to a change in the chemical properties of the soil and particularly of the soil solution.

The relationship between the soil–plant system and the grazing animal is complex and depends on a number of factors, including the proportion of grass in the animal's diet, other dietary trace element constituents, selectivity of grazing, digestibility of the diet and the form and availability of the ingested trace elements.

4.1. Soil Ingestion

Studies in New Zealand and Britain have demonstrated the importance of ingested soil to the dietary intake of trace elements by the grazing animal

(Healy, 1967, 1968). In Scotland ingested soil amounted to 14 % of the dry matter intake of sheep (Field and Purves, 1964). In more recent studies it has been shown that in winter months this figure may be as high as 40 % (Suttle *et al.*, 1975). Soil ingestion by cattle in Britain ranging from 1 to 10 % of the dry matter intake (140–1400 g soil/d) has been found over winter months in south-west England (Thornton, 1974). Where land is contaminated with copper, arsenic and lead—elements of low availability to pasture—cattle may ingest up to 10 times the amounts of the elements in the form of soil to that in herbage. It is therefore clear that on metal-contaminated soils, and for essential elements of low availability to the plant, such as cobalt and chromium and perhaps selenium, the soil–animal relationship may complement or in some cases override that of the soil–plant–animal relationship. However, little is as yet known about the availability of trace elements in ingested soil to the animal, the interactions between other dietary constituents and the metals, and the influence of microorganisms in the alimentary tract on metal solubility and absorption.

An example of the importance of soil ingestion as a major pathway of metal intake into the grazing animals is provided by a collaborative investigation conducted over the period 1975–76 in the mineralized district of Derbyshire between the Veterinary Investigation Service (part of the Ministry of Agriculture, Fisheries and Food), St Mary's Hospital Medical School and the Applied Geochemistry Research Group. In the early summer and autumn, blood samples were taken from 15 animals on 11 farms comprising background (100–200 ppm Pb), moderately contaminated (600–800 ppm Pb) and heavily contaminated (1000–2000 ppm Pb) locations, selected on the basis of the soil lead content. Blood lead levels quite clearly reflected lead in the soil, herd means in the early summer ranging from around 10 μg Pb/100 ml blood on the background farms to over 30 μg Pb/100 ml on high-lead farms. Values seldom exceeded 40 μg Pb/100 ml. Monthly sampling of pasture on these farms confirmed large seasonal effects, with peak concentrations in winter herbage (Fig. 1.4; Thornton and Kinniburgh, 1978). By the time animals were put out to graze in early May, lead in the pasture had fallen appreciably and differences between the three groups of farms were not significant over the grazing season. Pasture lead values over this period ranged from 10 to 40 ppm (dry matter) irrespective of the soil lead content. In spite of possible soil contamination, lead concentrations were mostly low. Soil ingestion in June (measured by Ti in faeces, taken at the same time as the blood samples) varied from 2·1 to 4·4 % of the dry matter intake. Based on an assumed dietary intake of 13·6 kg (30 lb) dry matter/d, lead intake ranged from around 30 mg Pb/d on control farms to over 1000 mg Pb/d on heavily contaminated land; from 10 to >80 % of the lead was present in ingested soil. Faecal lead increased both with total lead ingested and with soil lead content. Soil also constituted a source of zinc (3–36 % of the intake) and of copper (6–16 %) (Thornton and Kinniburgh, 1978). However, it is stressed that detailed studies are required to determine the relative solubility and absorption within the animal of trace metals ingested in soil and pasture herbage.

5. THE REGIONAL DISTRIBUTION OF TRACE METALS AND METHODS OF MAPPING

The growing awareness of the environmental significance of trace metals, the impact of deficiencies and excesses on agriculture, and potential hazards of metals in the food chain have highlighted the need for maps showing metal distribution on both a regional and national scale.

Ideally, trace metal maps for agricultural and environmental application would be based on the systematic sampling of soil and/or vegetation. Such maps have been compiled in New Zealand (Department of Scientific and Industrial Research, 1967), in parts of the US (Kubota and Allaway, 1972) and in Ireland (Brogan *et al.*, 1973), based on the careful selection of samples at relatively few locations and a large degree of geographical extrapolation. Rocks, soils, vegetation and crops have been analysed as part of a systematic wide scale geochemical survey of the State of Missouri (US Geological Survey, 1969–73), and similar samples have been used for the mapping of biogeochemical zones and provinces in the USSR (Kovalsky, 1970). In Britain this approach has so far proved impracticable because of the cost and time necessary to account for local differences owing to the complex geology and soil types. In spite of several *ad hoc* surveys undertaken in the course of soil survey and advisory work, there are few systematic data on either total or 'available' levels of trace metals in soils.

The need for maps showing metal distribution has been met in part by geochemical reconnaissance surveys undertaken in England and Wales by the Applied Geochemistry Research Group, Imperial College, and in Scotland by the Institute of Geological Sciences. These primary reconnaissance surveys are based on the systematic sampling of sediments taken from tributary drainage, a method based on the premise that active stream sediment represents a composite sample of the erosion products of rock and soil upstream from the sample site. These surveys are standard practice in mineral exploration. Exploratory work in Africa and the Republic of Ireland, and trial surveys over 7280 km^2 of England and Wales led to the recognition of broad-scale patterns of trace element distribution, reflecting both the natural composition of bedrock and soil and indicating specific geochemical facies within individual geological formations (Webb and Atkinson, 1965; Webb *et al.*, 1968; Nichol *et al.*, 1970*a*,*b*, 1971). At the same time, these geochemical maps draw attention to particular areas where land has been contaminated by past and present industrial activity (Thornton, 1975; Thornton and Webb, 1975). Similar maps have been prepared for parts of the Republic of Ireland (Kiely and Fleming, 1969) and geochemical reconnaissance maps have been published for the 13 000 km^2 of Northern Ireland (Webb *et al.*, 1973).

In 1969 nearly 50 000 stream sediment samples were taken from tributary drainage over England and Wales, representing an average sampling density of one sample/2·5 km^2. Analysis of the $<200\ \mu m$ fraction for Al, Ba, Ca, Co, Cr, Cu, Fe, Ga, K, Li, Mn, Ni, Pb, Sc, Sr, Sn and V was carried out using a 40-channel direct-reading emission spectrograph linked to an automatic card punch; Cd and Zn were determined by atomic absorption spectrometry and As and Mo by colorimetric procedures. Data have been presented by computer as grey-scale line-printer maps at scales ranging from 1 in:4 miles to 1:2 million and again as laser graphic plots printed on to diazo film. These maps have been on open-file at Imperial College since 1974 and have been presented in colour in *The Wolfson Geochemical Atlas of England and Wales* (Webb *et al.*, 1978).

It is stressed that from an agricultural/environmental view these maps do no more than focus attention on patterns of low and high trace metal distribution wherein to concentrate more detailed, time consuming and costly surveys based on soil and plant analysis. Only these subsequent detailed surveys can provide information on a local farm or field scale. Research at Imperial College over the past 15 years, aimed at the interpretation and use of these maps for agriculture, has confirmed relationships between regional geochemical data and the known distribution of agricultural disorders, and also has indicated further suspect areas requiring further investigation (Webb, 1964; Webb *et al.*, 1971; Thornton and Webb, 1970; Thomson *et al.*, 1972; Jordan *et al.*, 1975). The maps have delineated areas of low cobalt, copper, manganese and zinc; and high arsenic, cadmium, copper, lead, zinc, nickel, chromium and molybdenum. From a pollution viewpoint these geochemical surveys have been shown to provide a useful catalogue of baseline information related to the geochemistry of bedrock and soil parent materials; at the same time, the

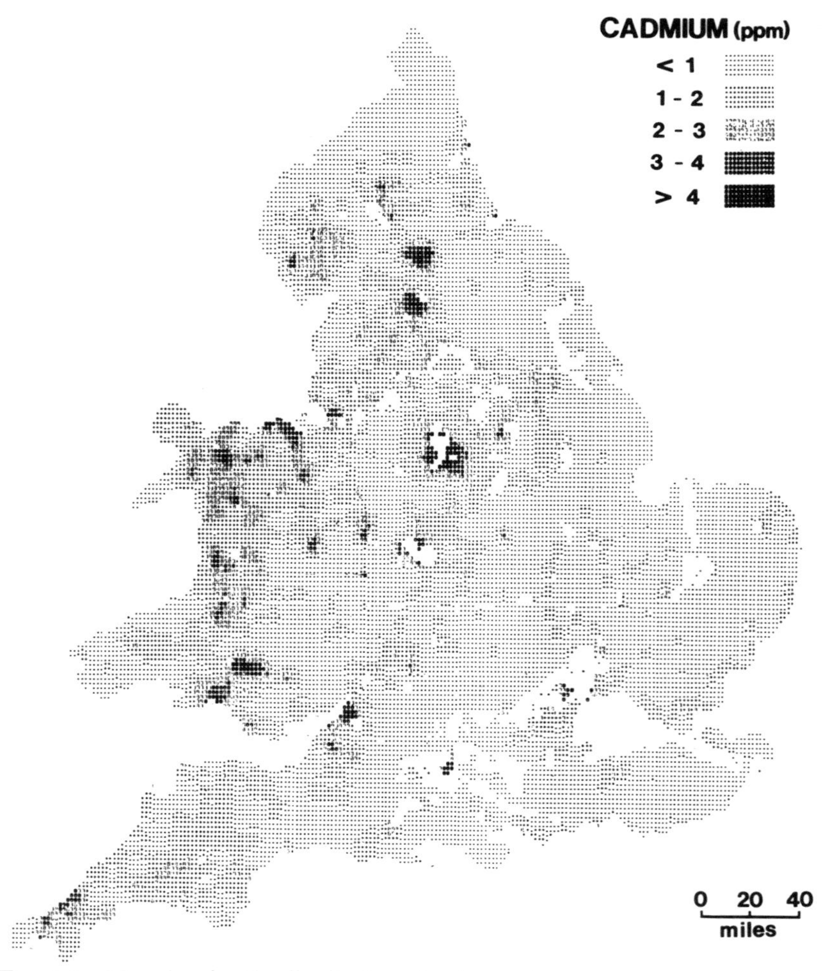

FIG. 1.5 Map showing the distribution of cadmium in stream sediment (compiled by the Applied Geochemistry Research Group as part of *The Wolfson Geochemical Atlas of England and Wales* (Webb *et al.*, 1978).

maps clearly show patterns reflecting metal contamination on a regional scale from past and present mining and smelting activities and from industrial and urban sources (Thornton, 1975; Thornton and Webb, 1975).

For example, the map for cadmium (Fig. 1.5) shows anomalies related to three main sources of the metal: (a) areas associated with sulphide mineralization contaminated by base-metal mining activity, including parts of south-west and central England and Wales; (b) areas contaminated by present-day and recent industrial processes including those near Bristol and the Swansea Valley affected by recent smelting of lead and zinc; (c) areas underlain by cadmium-rich marine shales of Carboniferous age (containing up to 100 ppm Cd) in which stream sediments and soils derived from the rock are enriched in the metal. Some soils in each of these types of anomaly have been shown to contain up to 20 ppm or more cadmium and present a potential hazard to the food chain. However, over 80% of England and Wales is represented by the lowest class of less than 1 ppm Cd in stream sediments. This corresponds to normal or background concentrations in rocks and soils.

It is of particular importance that multi-element geochemical data presented as a series of maps frequently draw attention to anomalous patterns of more than one metal in a specific area. It has already been emphasized that metals do not occur in isolation, whether the source is natural or man-made or a combination of both, and that synergistic and antagonistic interactions between metals are of prime significance in the soil–plant–animal system.

5. FUTURE RESEARCH REQUIREMENTS

As yet, little is known of the chemical and physical forms of trace metals and metalloids in soils and the ways in which these forms affect both their mobility in the soil and their availability, uptake and translocation by plants. Fractionation procedures referred to in this chapter are almost entirely restricted to the use of weak chemical extractants. The use of such methods provides a fund of empirical data which are often not applicable to practical situations under natural field conditions. Chemical tests of metal availability may be applicable to only a narrow range of soil types, and in almost every case no one test can be used on both non-calcareous and calcareous soils. The development of reliable methods of identifying and quantifying different forms of metals in soil and soil solution is an essential prerequisite to understanding the fate of natural and man-effected sources of metal in the soil and their potential significance to the food chain. The importance of time scale in relation to changes in the forms of metals in soils of different physical and chemical properties is another area requiring detailed study and is of particular significance to the impact of man-made inputs of metal to the soil–plant system.

Interactions between trace metals on the one hand and metals and major chemical components of the soil on the other are yet far from clearly understood and there is a need for more knowledge about relationships between metals, both synergistic and antagonistic, at all stages in the food chain.

In conclusion it should be mentioned that the role of soil microorganisms in trace metal geochemical processes has received little attention. It is suggested that the interface between soil chemistry and soil microbiology will prove to be an important future research area in seeking to understand the processes controlling trace metal behaviour in the soil–plant–animal system.

ACKNOWLEDGEMENTS

A number of examples are cited from the continuing programme of the Applied Geochemistry Research Group in environmental geochemistry, financed by the Agricultural Research Council and the Natural Environment Research Council.

REFERENCES

Allaway, W. H. (1977). Perspectives of molybdenum in soils and plants. In: *Molybdenum in the Environment*, Vol. 2 (W. R. Chappell and K. K. Petersen (eds)), Marcel Dekker, New York and Basel, pp. 317–39.

Alloway, B. J. and B. E. Davies (1971). Trace element content of soils affected by base metal mining in Wales. *Geoderma*. **5**: 197–207.

Andersson, A. and S. Nilsson (1972). Enrichment of trace elements from sewage sludge fertilizer in soils and plants. *Ambio*. **1**: 176–9.

Andersson, A. and S. Nilsson (1976). Influence on the levels of heavy metals in soil and plants from sewage sludge used as fertilizer. *Swedish J. agric. Res.* **6**: 151–9.

Antonovics, J. A. D., A. D. Bradshaw and A. G. Turner (1971). Heavy metal tolerance in plants. *Adv. Ecol. Res.* **7**: 1–85.

Archer, F. C. (1971). Factors affecting the trace element content of pastures. In: *Trace Elements in Soils and Crops*, Min. Ag. Fish. Fd. Tech. Bull. Vol. 21, HMSO, pp. 150–7.

Berggren, B. and S. Oden (1972). Analysresultat Rorande Tungmetaller Och Klorerade Kolväten 1 Rötslam Fran Svenska Reningsverk 1968–71. Institutionen für Markvetenskap Lantbrukshögskolan, Sweden.

Berrow, M. L. and J. C. Burridge (1980). Trace element levels in soils: effects of sewage sludge. In: *Inorganic Pollution and Agriculture*, Proc. A.D.A.S. Open Conference of Soil Scientists, London, 1977.

Berrow, M. L. and J. Webber (1972). Trace elements in sewage sludges. *J. Sci. Fd. Agric.* **23**: 93–100.

Bingham, F. T. (1963). Relation between phosphorus and micronutrients in plants. *Soil Sci. Soc. Am. Proc.* **27**: 389–91.

Blakeslee, P. A. (1973). Monitoring considerations for municipal waste water effluent and sludge application to land. US Environmental Protection Agency, US Dept. of Agriculture, Universities Workshop, July 1973, Champaign Urbana, Illinois.

Bloomfield, C. (1969). Mobilisation and fixation of iron and trace elements by aerobically decomposing plant matter. *Chem. Ind.* 1633–4.

Bloomfield, C., W. I. Kelso and G. Pruden. (1976). Reactions between metals and humidified organic matter. *J. Soil Sci.* **27**: 16–31.

Boswell, F. C. (1975). Municipal sewage sludge and selected element applications to soil: effect on soil and fescue. *J. Environ. Qual.* **4**: 267–73.

Brogan, J. C., G. A. Fleming and J. E. Byrne (1973). Molybdenum and copper in Irish pasture soils. *Irish J. Agric. Res.* **12**: 71–81.

Caldwell, T. H. (1971). Copper deficiency in crops. In: *Trace Elements in Soils and Crops*, Min. Ag. Fish. Fd. Tech. Bull. Vol. 21, HMSO, London, pp. 62–87.

Cannon, H. L. and B. M. Anderson (1971). The geochemist's involvement with pollution problems. In: *Environmental Geochemistry in Health and Disease* (H. L. Cannon and H. C. Hopps (eds)), Geological Society of America, Boulder, Colorado.

Cannon, H. L., G. G. Connally, J. B. Epstein, J. G. Parker, I. Thornton and B. G. Wixson (1978). Rocks: The geologic source of most trace elements. In: *Geochemistry and the Environment. Vol. III. Distribution of Trace Elements Related to the Occurrence of Certain Cancers, Cardiovascular Diseases and Urolithiasis.* National Academy of Sciences, Washington D.C., pp. 17–31.

Chappell, W. R. and K. K. Petersen (eds) (1977). *Molybdenum in the Environment*, Vol. I and II, Marcel Dekker, New York and Basel, 812 pp.

Chaudhry, F. M. and J. F. Loneragan (1970). Effects of nitrogen, copper and zinc nutrition of wheat plants. *Aust. J. agric. Res.* **21**: 865–79.

Colbourn, P., B. J. Alloway and I. Thornton (1975). Arsenic and heavy metals in soils associated with regional geochemical anomalies in south-west England. *Sci. Total Env.* **4**: 359–63.

Colbourn, P. and I. Thornton (1978). Lead pollution in agricultural soils. *J. Soil Sci.* **29**: 513–26.

Commission of the European Communities (1978). *Criteria (Dose/Effect Relationships) for Cadmium.* Pergamon Press, Oxford, 202 pp.

Davies, D. B., L. J. Hooper, R. R. Charlesworth, R. C. Little, C. Evans and B. Wilkinson (1971). Copper deficiency in crops. III. Copper disorders in cereals grown in chalk soils in south eastern and central southern England. In: *Trace Elements in Soils and Crops*, Min. Ag. Fish. Fd. Tech. Bull. Vol. 21, HMSO, London, pp. 88–118.

Davies, B. E. (1973). Occurrence and distribution of lead and other metals in two areas of unusual disease incidence in Britain. In: *Proceedings of the International Symposium on Environmental Health Aspects of Lead*, Commission of the European Communities, Luxembourg, pp. 125–34.

Davies, B. E. and L. J. Roberts (1975). Heavy metals in soils and radish in a mineralised limestone area of Wales, Great Britain. *Sci. Total Env.* **4**: 249–61.
Department of Scientific and Industrial Research (1967). *Soil Bureau Atlas.* D.S.I.R., Wellington, New Zealand.
Dowdy, R. H. and W. E. Larson (1975). The availability of sludge-borne metals to various vegetable crops. *J. Environ. Qual.* **4**: 278–82.
Ferguson, W. S., A. H. Lewis and S. J. Watson (1943). The teart pastures of Somerset. I. The cause of teartness. *J. Agric. Sci.* **33**: 44–51.
Field, A. C. and D. Purves (1964). The intake of soil by grazing sheep. *Proc. Nutr. Soc.* **23**: 24–5.
Fleming, G. A. (1962). Selenium in Irish soils and plants. *Soil Sci.* **94**: 28–35.
Fleming, G. A. (1965). Trace elements in plants with particular reference to pasture species. *Outlook Agric.* **4**: 270–85.
Fleming, G. A. (1968). Cobalt, selenium and molybdenum in Irish soils. Welsh Soils Disc. Gp. No. 9, pp. 41–56.
Fleming, G. A. and J. Delaney (1961). Copper and nitrogen in the nutrition of wheat on cutaway peat. *Irish J. Agric. Res.* **1**: 81–2.
Fleming, G. A. and T. Walsh (1957). Selenium occurrence in certain Irish soils and its toxic effect on animals. *Roy. Irish. Acad. Proc.* **58**: (Sect. B) 151–66.
Fletcher, W. E. (1968). Geochemical reconnaissance in relation to copper deficiency in livestock in the Southern Pennines. Unpublished Ph.D. Thesis, University of London.
Fulkerson, W. (1975). Cadmium—the dissipated element—revisited. Oak Ridge National Laboratory, Tennessee.
Furr, A. K., A. W. Laurence, S. S. C. Tong, M. C. Grandolfo, R. A. Hofstader, C. A. Bache, W. H. Guttenham and D. J. Lisk (1976). Multi-element and chlorinated hydrocarbon analysis of municipal sewage sludge of American cities. *Environ. Sci. Technol.* **10**: 683–7.
Gray, K. R. and A. J. Biddlestone (1980). Agricultural use of composted town refuse. In: *Inorganic Pollution and Agriculture*, Proc. A.D.A.S. Open Conference of Soil Scientists, London, 1977.
Hawkes, H. E. and J. S. Webb (1962). *Geochemistry in Mineral Exploration.* Harper and Row, New York and Evanston. 415 pp.
Healy, W. B. (1967). Ingestion of soil by sheep. *Proc. N.Z. Soc. Animal Production.* **27**: 109–20.
Healy, W. B. (1968). Ingestion of soil by dairy cows. *N.Z.J. Agric. Res.* **11**: 487–99.
Henkens, C. H. (1957). Copper in arable land. *Landbouwvoorlichting.* **14**: 581–9.
Holmes, R. (1975). The regional distribution of cadmium in England and Wales. Unpublished Ph.D. Thesis, University of London.
Ireland, M. P. (1975). The effect of the earthworm *Dendrobaena rubida* on the solubility of lead, zinc and calcium in heavy metal contaminated soil in Wales. *J. Soil Sci.* **26**: 313–18.
Jarvis, S. C., L. H. P. Jones and M. J. Hopper (1976). Cadmium uptake from solution by plants and its transport from roots to shoots. *Plant Soil.* **44**: 179–91.
Jenne, E. A. (1977). Trace element sorption by sediments and soils—sites and processes. In: *Molybdenum in the Environment*, Vol. 2 (W. R. Chappell and K. K. Petersen (eds)), Marcel Dekker, New York and Basel, pp. 425–553.

Jennett, J. C., B. J. Wixson, E. Bolter, I. H. Lowsley, D. D. Hemphill, W. H. Tranter, N. L. Gale and K. Purushotaman (1977). Transport and distribution around mines, mills and smelters. In: *Lead in the Environment*, Report NSF/RA-770214, pp. 135–78.

Jones, L. H. P. and C. R. Clement (1972). Lead uptake by plants and its significance for animals. In: *Lead in the Environment*, Institute of Petroleum, London, pp. 29–33.

Jordan, W. J. (1975). The application of regional geochemical reconnaissance to arable cropping in England and Wales. Unpublished Ph.D. Thesis, University of London.

Jordan, W. J., B. J. Alloway and I. Thornton (1975). The application of regional geochemical reconnaissance data in areas of arable cropping. *J. Sci. Fd. Agric.* **26**: 1413–24.

Keeley, H. C. M. (1972). Copper, cobalt and manganese in relation to geochemical reconnaissance and agriculture. Unpublished Ph.D. Thesis, University of London.

Kiely, P. V. and G. A. Fleming (1969). Geochemical survey of Ireland: Meath–Dublin area. *Proc. Roy. Irish Acad. B.* **68**: p. 28.

Kovalsky, V. V. (1970). The geochemical ecology of organisms under conditions of varying contents of trace elements in the environment. In: *Trace Element Metabolism in Animals* (C. F. Mills (ed)), Proc. WAAP/IBP International Symposium, E. & S. Livingstone, London, pp. 385–96.

Kubota, J. (1977). Molybdenum status of United States soils and plants. In: *Molybdenum in the Environment*, Vol. 2 (W. R. Chappell and K. K. Petersen (eds)), Marcel Dekker, New York and Basel, pp. 555–81.

Kubota, J. and W. H. Allaway (1972). Geographic distribution of trace element problems. In: *Micronutrients in Agriculture* (J. J. Mortredt, P. M. Gordiano and W. L. Lindsay (eds)), Soil Science Soc. Amer., Madison, Wisconsin, pp. 525–54.

Lagerwerff, J. V. (1972). Lead, mercury and cadmium as environmental contaminants. In: *Micronutrients in Agriculture* (J. J. Mortredt, P. M. Gordiano and W. L. Lindsay (eds)), Soil Sci. Soc. Amer., Madison, Wisconsin, pp. 593–636.

Le Riche, H. H. (1959). Molybdenum in the Lower Lias of England in relation to the incidence of teart. *J. Soil Sci.* **10**: 133–6.

Le Riche, H. H. (1968). Metal contamination of soil in the Woburn market-garden experiment resulting from the application of sewage. *J. Agric. Sci. Camb.* **71**: 205–8.

Levinson, A. A. (1974). *Introduction to Exploration Geochemistry.* Applied Publishing Ltd, Calgary, 611 pp.

Lewis, A. H. (1943). The teart pastures of Somerset. II. The relation between soil and teartness. *J. Agric. Sci. Camb.* **33**: 52–7.

Lindsay, W. L. (1972*a*). Inorganic phase equilibria of micronutrients in soils. In: *Micronutrients in Agriculture* (J. J. Mortredt, P. M. Gordiano and W. L. Lindsay (eds)), Soil Sci. Soc. Amer., Madison, Wisconsin, pp. 41–57.

Lindsay, W. L. (1972*b*). Zinc in soils and plant nutrition. *Adv. Agron.* **24**: 147–84.

Marples, A. E. (1979). The occurrence and behaviour of cadmium in soils and its uptake by pasture grasses in industrially contaminated and naturally metal-rich environments. Unpublished Ph.D. Thesis, University of London.

Marples, A. E. and I. Thornton (1980). The distribution of cadmium derived from geochemical and industrial sources in agricultural soils and pasture herbage in parts of Britain. In: *Proceedings of the Second International Cadmium Conference*, Cannes, 1979.

Mengel, K. and E. A. Kirkby (1978). Copper. In: *Principles of Plant Nutrition*, Ch. 16, International Potash Institute, Wurblanfen-Bern, pp. 463–74.

Mercer, E. R. and J. L. Richmond (1970). Fate of nutrients in soil: Copper. Letcombe Laboratory Annual Report No. 9, Letcombe Laboratory, Wantage, Oxfordshire.

Mitchell, R. L. (1964). Trace elements in soils. In: *Chemistry of the Soil*, 2nd Ed (F. E. Bear (ed)), Reinhold Publ. Co., New York, pp. 320–68.

Mitchell, R. L. (1971). Trace elements in soils. In: *Trace Elements in Soils and Crops*, Min. Agric. Fish. Fd. Tech. Bull. Vol. 21, HMSO, London, pp. 8–20.

Mitchell, R. L. (1974). Trace element problems on Scottish soils. *Neth. J. Agric. Sci.* **22**: 295–304.

Mitchell, R. L. and J. W. S. Reith (1966). The lead content of pasture herbage. *J. Sci. Fd. Agric.* **17**: 437–40.

Nichol, I., I. Thornton, J. S. Webb, W. K. Fletcher, R. F. Horsnail, J. Khaleelee and D. Taylor (1970*a*). Regional geochemical reconnaissance of the Derbyshire area. Rep. No. 70/2, Inst. Geol. Sci., London.

Nichol, I., I. Thornton, J. S. Webb, W. K. Fletcher, R. F. Horsnail, J. Khaleelee and D. Taylor (1970*b*). Regional geochemical reconnaissance of the Denbighshire area, Rep. No. 70/8, Inst. Geol. Sci., London.

Nichol, I., I. Thornton, J. S. Webb, W. K. Fletcher, R. F. Horsnail and J. Khaleelee (1971). Regional geochemical reconnaissance of part of Devon and north Cornwall, Rep. No. 71/2, Inst. Geol. Sci., London.

Nye, S. M. and P. J. Peterson (1975). The content and distribution of selenium in soils and plants from seleniferous areas in Eire and England. In: *Trace Substances in Environmental Health—X* (D. D. Hemphill (ed)), University of Missouri, Columbia, Missouri, pp. 113–21.

Page, A. L. (1974). Fate and effects of trace elements in sewage sludge when applied to agricultural lands. Environ. Prot. Technol. Ser. EPA-670/2-74-005, US Environmental Protection Agency, Ohio.

Purves, D. (1977). *Trace Element Contamination of the Environment*. Elsevier Scientific Publ. Co., Amsterdam, 260 pp.

Purves, D. and E. J. Mackenzie (1973). Effects of applications of municipal compost on uptake of copper, zinc and boron by garden vegetables. *Plant Soil.* **39**: 361–71.

Rai, D. and W. L. Lindsay (1975). A thermodynamic model for predicting the formation, stability and weathering of common soil minerals. *Soil Sci. Soc. Am. Proc.* **39**: 991–6.

Runnells, D. D., D. S. Kaback and E. M. Thurman (1977). Geochemistry and sampling of molybdenum in sediments, soils and plants in Colorado. In: *Molybdenum in the Environment*, Vol. 2 (W. R. Chappell and K. K. Petersen (eds)), Marcel Dekker, New York and Basel, pp. 387–423.

Russel, F. C. and D. L. Duncan (1956). *Minerals in Pasture: Deficiencies and excesses in Relation to Animal Health*, Animal Nutr. Tech. Commun. No. 15, 2nd Ed, Commonwealth Bureau, London.

Siegel, F. R. (1974). *Applied Geochemistry*. John Wiley, New York, 353 pp.

Stenstrom, T. and M. Walter (1974). Cadmium and lead in Swedish commercial fertilizers. *Ambio*. **3**: 91.

Suttle, N. F., B. J. Alloway and I. Thornton (1975). An effect of soil ingestion on the utilization of dietary copper by sheep. *J. Agric. Sci. Camb.* **84**: 249–54.

Swaine, D. J. (1962). *The Trace Element Content of Fertilizers*, Commonw. Bur. Soil Tech. Commun. No. 52, Commonwealth Agricultural Bureau, Farnham Royal, 306 pp.

Swaine, D. J. (1977). Trace elements in fly ash. In: *Geochemistry 1977*, DSIR Bull. 218, Wellington, New Zealand, pp. 127–31.

Swaine, D. J. and R. L. Mitchell (1960). Trace element distribution in soil profiles. *J. Soil Sci.* **11**: 347–68.

Szalay, A. (1964). Cation exchange properties of humic acids and their importance in the geochemical enrichment of UO_2^{++} and other cations. *Geochim. Cosmochim. Acta*. **28**: 1605–14.

Szalay, A. (1969). Accumulation of uranium and other micro-metals in coal and organic shales and the role of humic acids in their geochemical enrichment. *Arkiv. Mineraologi Geologi.* **5**: 23–36.

Thomson, I. (1971). Regional geochemical studies of black shale facies with particular reference to trace element disorders in animals. Unpublished Ph.D. Thesis, University of London.

Thomson, I., I. Thornton and J. S. Webb (1972). Molybdenum in black shales and the incidence of bovine hypocuprosis. *J. Sci. Fd. Agric.* **23**: 871–91.

Thoresby, P. and I. Thornton (1980). Heavy metals and arsenic in soil, pasture herbage and barley in some mineralised areas in Britain: Significance to animal and human health. In: *Trace Substances in Environmental Health—XIII* (D. D. Hemphill (ed)), Columbia, Missouri, University of Missouri.

Thornton, I. (1968). The application of regional geochemical reconnaissance to agricultural problems. Unpublished Ph.D. Thesis, University of London.

Thornton, I. (1974). Biogeochemical and soil ingestion studies in relation to trace element nutrition of livestock. In: *Trace Element Metabolism in Animals—2* (W. G. Hoekstra *et al.* (eds)), University Park Press, Baltimore, pp. 451–4.

Thornton, I. (1975). Applied geochemistry in relation to mining and the environment. In: *Minerals and the Environment* (M. J. Jones (ed)), Institution of Mining and Metallurgy, London, pp. 87–102.

Thornton, I. (1977). Biogeochemical studies on molybdenum in the United Kingdom. In: *Molybdenum in the Environment*, Vol 2 (W. R. Chappell and K. K. Petersen (eds)), Marcel Dekker, New York and Basel, pp. 341–69.

Thornton, I., P. Abrahams and H. Matthews (1979). Some examples of the environmental significance of heavy metal anomalies disclosed by *The Wolfson Geochemical Atlas of England and Wales*. In: *Proceedings Int. Conf. Management and Control of Heavy Metals in the Environment*, London, pp. 218–21.

Thornton, I. and D. G. Kinniburgh (1978). Intake of lead, copper and zinc by cattle from soil and pasture. In: *Trace Element Metabolism in Man and Animals—3* (M. Kirchgessner (ed)), Institut für Ernahrungsphysiologie, Technische Universität München Freisung—Weihenstephan, p. 499.

Thornton, I. and J. S. Webb (1970). Geochemical reconnaissance and the detection of trace element disorders in animals. In: *Trace Element Metabolism in Animals, Proc. WAAP/IBP International Symposium* (C. F. Mills (ed)), E. & S. Livingstone, London, pp. 397–407.

Thornton, I. and J. S. Webb (1975). Trace elements in soils and surface waters contaminated by past metalliferous mining in parts of England. In: *Trace Substances in Environmental Health—IX* (D. D. Hemphill (ed)), University of Missouri, Columbia, Missouri, pp. 77–88.

Thornton, I. and J. S. Webb (1976). Distribution and origin of copper deficient and molybdeniferous soils in the United Kingdom. Proc. Copper in Farming Symposium, Copper Development Association, London.

Thornton, I. and J. S. Webb (1980*a*). Regional distribution of trace element problems in Great Britain. In: *Applied Soil Trace Elements* (B. E. Davies (ed)), John Wiley, Chichester, pp. 382–439.

Thornton, I. and J. S. Webb (1980*b*). Trace elements in soils and plants. In: *Food Chains and Human Nutrition* (K. L. Blaxter (ed)), Applied Science Publishers Ltd, London, pp. 273–312.

Underwood, E. J. (1966). *The Mineral Nutrition of Livestock*. Commonwealth Agricultural Bureau and F.A.O. 237 pp.

Underwood, E. J. (1971). *Trace Elements in Human and Animal Nutrition*, 3rd Ed, Academic Press, New York and London. 543 pp.

United States Geological Survey (1969–73). Geochemical Survey of Missouri, Open-file Reports. US Geological Survey, Denver, Colorado.

Unwin, R. J. (1980). Copper in pig slurry: Some effects and consequences of spreading on grassland. In: *Inorganic Pollution and Agriculture*, Proc. A.D.A.S. Open Conference of Soil Scientists, London, 1977.

Viets, F. C., L. C. Boawn and C. L. Crawford (1954). Zinc contents and deficiency symptoms of 26 crops grown on a zinc deficient soil. *Soil Sci.* **78**: 305–16.

Vlek, P. L. G. and W. L. Lindsay (1977). Molybdenum contamination in Colorado pasture soils. In: *Molybdenum in the Environment*, Vol. 2 (W. R. Chappell and K. K. Petersen (ed)), Marcel Dekker, New York and Basel, pp. 619–50.

Walsh, T., M. Neenan and L. B. O'Moore (1953). High molybdenum levels in herbage on acid soils. *Nature, Lond.* **171**: p. 1120.

Webb, J. S. (1964). Geochemistry and life. *New Scientist*. **23**: 504–7.

Webb, J. S. and W. J. Atkinson (1965). Regional geochemical reconnaissance applied to some agricultural problems in Co. Limerick, Eire. *Nature, Lond.* **208**: 1056–9.

Webb, J. S., P. L. Lowenstein, R. J. Howarth, I. Nichol and R. Foster (1973). *Provisional Geochemical Atlas of Northern Ireland.* Applied Geochemistry Research group Tech. Comm. No. 60.

Webb, J. S., I. Nichol and I. Thornton (1968). The broadening scope of regional geochemical reconnaissance. *XXIII International Geological Cong.* **6**: 131–47.

Webb, J. S., I. Thornton, R. J. Howarth, M. Thompson and P. L. Lowenstein (1978). *The Wolfson Geochemical Atlas of England and Wales.* Oxford University Press, Oxford, 69 pp.

Webb, J. S., I. Thornton and I. Nichol (1971). The agricultural significance of regional geochemical reconnaissance in the United Kingdom. In: *Trace Elements in Soils and Crops*, Min. Agric. Fish. Fd. Tech. Bull. Vol. 21, HMSO, London, pp. 1–7.

Williams, C. H. and D. G. David (1973). The effect of superphosphate on the cadmium content of soils and plants. *Aust. J. Soil Res.* **11**: p. 43.

Williams, C. and I. Thornton (1972). The effect of soil additives on the uptake of molybdenum and selenium from soils from different environments. *Plant Soil.* **36**: 395–406.

Wood, P. (1975). Regional geochemical studies in relation to agriculture in areas underlain by sandstones. Unpublished Ph.D. Thesis, University of London.

CHAPTER 2

Sources of Metal and Elemental Contamination of Terrestrial Environments

B. FREEDMAN

Department of Biology and Institute for Resource and Environmental Studies, Dalhousie University, Canada

and

T. C. HUTCHINSON

Department of Botany and Institute for Environmental Studies, University of Toronto, Canada

1. INTRODUCTION

Heavy metals, as environmental contaminants of terrestrial ecosystems, are not a recent phenomenon. They are ubiquitous in trace concentrations in soils and vegetation, and in fact many are required by plants and animals as micronutrients. In addition, naturally occurring surface mineralizations can produce metal concentrations in soils and vegetation that are as high, or higher, than those found around man-made sources. The searches by prospectors for such metal deposits, frequently using biogeochemical techniques, are a part of our history. These cases of natural 'pollution' are, however, relatively localized, and it was not until the recent era of industrialization that widespread contaminations by heavy metals (and other classes of pollutants as well) have occurred.

The man-made sources of metal contamination of terrestrial environments are mainly associated with certain industrial activities, particularly those of the metal smelting and refining industries. Some contamination also occurs through agricultural practices, from automobile emissions, from the disposal of metal-containing mining wastes, and from atmospheric emissions from coal-fired generating plants and municipal incinerators. These various sources, and their environmental impacts in causing metal contamination, are the topic of the present chapter.

2. BACKGROUND METAL CONCENTRATIONS IN SOILS AND VEGETATION

Several recent sources in the literature list typical analyses of rocks, soils and vegetation collected from uncontaminated areas (e.g. Bowen (1966) and Aubert and Pinta (1977)). These are useful for comparison with data from metal-contaminated areas. Tables 2.1 and 2.2 list such data for soils and vegetation, respectively.

TABLE 2.1
ELEMENTAL COMPOSITION OF TYPICAL UNCONTAMINATED SOILS. (AFTER BOWEN, 1966)

Metal	Mean concentration (ppm (dw))	Range (ppm (dw))
Ag	0·1	0·01–5
Al	71 000	10 000–300 000
As	6	0·1–40
Cd	0·06	0·01–0·7
Co	8	1–40
Cr	100	5–3 000
Cu	20	2–100
Fe	38 000	7 000–550 000
Hg	0·03	0·01–0·3
Mn	850	100–4 000
Mo	2	0·2–5
Ni	40	10–1 000
Pb	10	2–200
Sn	10	2–200
Ti	5 000	1 000–10 000
V	100	20–500
Zn	50	10–300

TABLE 2.2

ELEMENTAL COMPOSITION OF UNCONTAMINATED PLANT TISSUES (ppm (dw)). (AFTER BOWEN, 1966 AND ALLAWAY, 1968)

Metal	Lichens and fungi	Bryophytes	Pteridophytes	Gymnosperms	Angiosperms	Range in agricultural crop plants
Ag	0·15	0·1	0·23	0·07	0·06	—
Al	29	1 400	—	65	550	—
As	—	—	—	—	0·2	0·1–5·0
Cd	4	0·1	0·5	0·24	0·64	0·2–0·8
Co	0·5	0·33	0·8	0·2	0·48	0·05–0·5
Cr	1·5	2	0·8	0·16	0·23	0·2–1·0
Cu	15	7	15	15	14	4–15
Fe	130	1 200	300	130	140	—
Hg	—	—	—	—	0·015	—
Mn	25	290	250	330	630	15–100
Mo	1·5	0·7	0·8	0·13	0·9	1–100
Ni	1·5	2·5	1·5	1·8	2·7	1
Pb	50	3·3	2·3	1·8	2·7	0·1–10
V	0·67	2·3	13	0·69	1·6	0·1–10
Zn	150	50	77	26	160	15–200

3. SOURCES OF METAL CONTAMINATION

As considered here, there are seven major categories of sources of metal contamination of the terrestrial environment. These are:

1. Natural sources, such as surface mineralizations, volcanic outgassings, spontaneous combustions or forest fires.
2. The use of metal-containing agricultural sprays or soil amendments.
3. The disposal of wastes from mines or mills.
4. Emissions from large industrial sources, such as metal smelters and refineries.
5. Emissions from municipal utilities, such as coal- or oil-fired electricity generating stations or municipal incinerators.
6. Emissions from moving sources, principally automobiles.
7. Other relatively minor sources of terrestrial contamination, such as smaller scale industries that process metals.

The impacts of these various sources in terms of resultant contamination of soils and vegetation are discussed in detail below.

3.1. Naturally-occurring Contamination

3.1.1. Surface mineralizations

The occurrence of surface or near-surface metal mineralizations frequently gives rise to relatively localized zones of contamination. This existence of anomalously high metal concentrations in soils, vegetation or lake or stream waters, and sediments overlying economically significant metal deposits, has given rise to the prospecting technique known as *biogeochemical prospecting*. Some specific examples of the use of this technique are described in a number of recent publications (e.g. Cannon, 1960; Malyuga, 1964; Warren *et al.*, 1966*a*,*b*; Allan, 1971; Forgeron, 1971; Wolfe, 1971; Webb *et al.*, 1973; Leland *et al.*, 1974; Thornton, 1975 and Brooks *et al.*, 1977). In one particularly elegant application of this technique, geochemical atlases illustrating the regional distribution of 26 elements have been prepared for Northern Ireland, England and Wales (Webb *et al.*, 1973; Thornton, 1975).

In some cases, the metal contamination overlying surface metal mineralizations can be quite severe, and the metal concentrations in soils and vegetation may be comparable in magnitude to some of the worst examples of man-effected pollution. For example, Boyle (1971) found copper concentrations of up to 10% (dw) (dry weight) in surface peat samples at a site called the Tantramar Copper Swamp in New Brunswick, Canada. At the same site, Stone and Timmer (1975) found copper concentrations of 2·05% in surface peats (9–15 cm), and 5·2% at lower depths (15–30 cm). The source of contamination was artesian spring water having high copper concentrations. Similarly, Forgeron (1971) describes soil concentrations of up to 3% lead plus zinc at a site on Baffin Island, in the Canadian Arctic Archipelago. In another example, Warren *et al.* (1966*a*) found mercury contents ranging from 1 to 10 ppm (dw) in soils overlying cinnabar (HgS) deposits in British Columbia, Canada, compared with *c.* 0·1 ppm at uncontaminated sites.

As one might expect, the severe soil contamination at some of these metalliferous sites has led to the existence of very high metal concentrations in vegetation. This phenomenon is particularly acute in genetically adapted hyperaccumulator populations or species, which may be endemic to such locations. For example, nickel contents of up to 10% (dw) have been found in *Alyssum bertolonii* and *A. murale* (Mishra and Kar, 1974 (citing Malyuga, 1964)). In the New Caledonian plant, *Sebertia acuminata*, nickel contents of up to 25% (dw) have been analysed in latex (Jaffre *et al.*, 1976). In a chemical survey of 181 herbarium specimens comprising 70 *Rhinorea* species, Brooks *et al.* (1977) found specimens of *R. bengalensis* containing up to 1·75% (dw), and *R. javanica* having up to 2170 ppm (dw) of nickel.

Similarly, copper indicators, and even copper endemics, have been described. The labiate *Becium homblei* occurs on copper deposits in Zaire, Zimbabwe and Zambia. Cannon (1960) stated that this species was responsible for the discovery of copper deposits in Zambia and Zimbabwe, where its occurrence was believed to be confined to soils with >1000 ppm (dw) copper. Reilly (1967), and Reilly and Reilly (1973) described *B. homblei* as a cuprophile, tolerant to greater than 70 000 ppm (dw) copper in soil, and having up to 17% of the copper in the leaves being organically bound to the cell walls.

Copper mosses were originally described from Scandinavia, and later from Alaska (Persson, 1948; Shacklette, 1965). More recently, Alstrup and Hamen (1977) reported three species of lichen (*Stereocaulon paschale*, *Umbilicaria lyngei* and *Lecanora polytropa*) on outcrops containing 2% copper on Disko Island, Greenland.

In a further example, Reay (1972) described arsenic enrichment in a number of aquatic plants growing at New Zealand springwater sites contaminated with arsenic from geothermal sources. Arsenic concentrations of up to 970 ppm (dw) were analysed in *Ceratophyllum demersum*, compared with 1·4 ppm (dw) in non-enriched waters.

One interesting soil type, derived in part from serpentine minerals, is a particularly well-studied example of naturally occurring metal contamination, and is described in some detail here. These soils are derived from ultramafic minerals, particularly olivine, orthorhombic and monoclinic pyroxenes and hornblende; also from the secondary alteration products of these minerals, particularly the serpentine group minerals, fibrous amphiboles and talc (Aumento, 1970; Douglas, 1970; Proctor and Woodell, 1975). The serpentine minerals are a group of layered silicates having the general formula $Mg_6Si_4O_{10}(OH)_8$, in which a significant substitution for magnesium by other elements, especially nickel, cobalt and iron can occur. Nickeliferous pyrite and chromite may also be present in significant quantities in some serpentine-derived soils. In many cases, these soils are inhibitory to the growth of non-adapted vegetation. This characteristic is generally ascribed to a number of chemical edaphic factors, especially the (usually) high contents of nickel, chromium and cobalt, and the low contents of available calcium, potassium, phosphorus, nitrogen and molybdenum. Magnesium and iron are also present in high concentration, but these are not thought to be directly toxic to vegetation on these sites (Whittaker, 1954; Proctor and Woodell, 1975). Striking physiognomic changes occur in the vegetation on serpentine soils. The serpentine flora is generally low-growing and stunted. In one Cuban site, for example, the serpentine community has the appearance of park-like savannah in an area of tropical rain forest.

Various authors have described heavy metal contaminations of soils and vegetation growing on serpentine sites. Proctor and Woodell (1975), in a comprehensive review of the ecology of soils derived from serpentine, noted that nickel contents of the order of thousands of ppm are common, with some sites having up to 25 000 ppm (dw). Peterson (1975) also reviewed much of the literature on the chemical ecology of serpentine-derived soils, and cited soil analyses from a Rhodesian site where up to 125 000 ppm (dw) of chromium and 5500 ppm (dw) of nickel were observed. As might be expected, some of these soil metals are available for plant uptake, and plant nickel concentrations of from 600 to 10 000 ppm (dw) in ash have been observed, compared with 20–70 ppm in control plants (Proctor and Woodell, 1975). In fact, the toxicities of some poor agricultural soils in Scotland and Rhodesia have been ascribed to direct nickel or chromium toxicities (Hunter and Vergnano, 1952; Vergnano and Hunter, 1952; Soane and Saunder, 1959).

Sites having soils derived from serpentine often contain unique and biogeographically interesting natural plant communities (reviewed in Proctor and Woodell, 1975). These sites also have a high percentage of endemics and ecotypes. On more productive sites, the serpentine species and ecotypes would be quickly eliminated through competition with species that are better adapted to more moderate habitat conditions. Scoggan (1950) identified four groups of plants (apart from common elements of the local flora) that find refugia on North American serpentine barrens. These were: (1) Arctic circumpolar species; (2) Arctic–American species, (3) disjunct American species, and (4) local serpentine endemics. A notable example of the latter group is the serpentine-indicating fern *Cheilanthes siliquosa*. None of these four groups of plants can exist locally away from serpentine outcrops. Not surprisingly, nickel-tolerant ecotypes have been identified from various serpentine sites in North America and Europe (e.g. Kruckeberg, 1954; Proctor, 1971*a*,*b*; Proctor and Woodell, 1975).

3.1.2. Other natural sources of contamination

A number of other natural sources of metal contamination exist. However, because of a paucity of emission data, it is difficult to assess their significance. It is well recognized, for instance, that volcanic outgassings are major natural emitters of such gaseous pollutants as sulphur dioxide and hydrogen sulphide (e.g. Kellogg *et al.*, 1972; Stoiber, 1973; Granat *et al.*,

1975). Recent evidence indicates that these sources may also contribute metals to the atmosphere, particularly the relatively volatile elements. Siegel and Siegel (1978) monitored atmospheric mercury at various locales in Hawaii during a period of volcanic eruption in 1977. They found large increases in atmospheric mercury during periods of intense activity, with concentrations ranging from 50 to $>200\,\mu g/m^3$, compared with background values of $<1\,\mu g/m^3$.

Other interesting natural sources of metal contamination, although of relatively localized significance, are spontaneous combustions of carbonaceous materials. Hutchinson *et al.* (1978) reported data from a tundra site near Cape Bathurst, in the western Canadian Arctic, which was being fumigated by spontaneous combustions of sulphide-rich bituminous shales that had been exposed by the erosion of 50-m high sea cliffs. Although the principal pollutants at this site (appropriately named the 'Smoking Hills' by early explorers) were sulphur dioxide and hydrogen sulphide, increases in atmospheric concentrations of several volatile elements were also found (Table 2.3). In addition, elevated concentrations of arsenic and vanadium occurred in the frequently fumigated soils, where concentrations were *c.* 2–3 times higher than those found in tundra soils, presumably as a result of atmospheric deposition. These soils were highly acidic, as were the waters of thermokarst tundra ponds in the fumigated zone, with soil pH's as low as 2·0, and water pH's as low as 1·8 being measured. This acidification was largely attributed to the long-term dry deposition of sulphur dioxide. At

TABLE 2.3

TOTAL SUSPENDED PARTICULATES AND AEROSOLS IN AIR AT SMOKING HILLS, NORTHWEST TERRITORIES, CANADA, SAMPLED DURING GROWING SEASON OF 1976. ANDERSON HI-VOL SAMPLER (ng/m^3, 24 h AVERAGE). (AFTER HUTCHINSON *et al.* (1978))

Element	Control (0 % fumigated)	10 % Fumigation	100 % Fumigation (a)	100 % Fumigation (b)
As	0·2	8·1	12·5	36·0
Br	4·8	16·5	54·5	73·0
Cl	593	221	243	89
Mn	2·1	0·6	53·3	108·7
Na	406	171	54	67
S	<300	3 000	120 000	155 000
Sb	0·5	0·6	1·0	1·2
Se	<31	<42	128	269
V	0·6	0·2	0·4	2·9

these low pH's, large quantities of heavy metals were mobilized from the watershed soils or sediments of these ponds, resulting in very high concentrations of various heavy metals in solution (Table 2.4). This constitutes a naturally occurring phenomenon that is analogous to the solubilization of soil metals by acid mine drainage solutions and from acid shales, to be discussed later.

TABLE 2.4

CHEMICAL ANALYSIS OF FILTERED WATER FROM TUNDRA PONDS IN THE VICINITY OF THE SMOKING HILLS, NORTHWEST TERRITORIES, CANADA. GEOMETRIC MEANS OF DATA (mg/litre). (AFTER HUTCHINSON *et al.* (1978))

pH range	Al	Fe	Mn	Zn	Ni	Cd	As	Number
1·8–2·5	270	500	61	14	6·3	0·52	0·130	4
2·5–3·5	5·5	18	15	0·45	0·21	0·022	0·005	14
3·5–4·5	1·1	1·2	3·6	0·12	0·04	0·011	0·005	9
4·5–5·5	<0·6	0·5	23	0·03	0·04	0·001	0·004	1
5·5–6·5	<0·2	0·2	1·8	0·05	0·06	0·012	0·006	1
6·5–7·5	<0·8	<0·04	0·7	0·08	0·02	0·001	0·004	4
7·5–8·5	<0·7	0·1	<0·5	0·04	0·004	0·003	0·005	8

Other significant natural sources of air pollutants, such as forest fires, have not yet received attention as to their possible roles as sources of heavy metal particulates. Several recent studies have shown that forest fires may contribute considerable quantities of carbonaceous particulate matter to the atmosphere, as well as large quantities of ozone and nitrogen dioxide (Evans *et al.*, 1974; Anon, 1976*a*, 1978*a*; Radke *et al.*, 1977). Forest fires consumed an estimated $8{\cdot}5 \times 10^5$ ha in 1974 in Canada, and in the process contributed some 275×10^3 MT/year of particulates to the atmosphere. These particulate emissions, plus another 55×10^3 MT/yr emitted as a result of controlled slash burning following clear-cutting, accounted for *c.* 15% of the total Canadian particulate emissions for that year (Anon, 1978*a*). Similarly, an estimated $0{\cdot}5 \times 10^6$–50×10^6 MT/yr of particulates were calculated to be emitted to the atmosphere by forest fires in the US in recent years, with the range of emissions largely reflecting variations in the amounts of forest lands burning (Radke *et al.*, 1977). We are not aware of any studies that have specifically studied metal emissions from forest fires. One study, however, reported on emissions of major cations (i.e. Ca, Mg, K) from forest fires, and found the '. . . amounts of major cations released by [simulated] light litter fires cannot be significant in comparison with the amounts of ions normally present in rain' (Lewis, 1975). Assuming that metals behave similarly to the major cations, it thus would appear that forest fires may not be major contributors to the atmospheric load of heavy metals, in spite of their tremendous particulate outputs. However, it can be expected that highly volatile elements, such as arsenic, mercury, selenium and cadmium will be lost from both burning vegetation and also from hot surface soils.

3.2. Agricultural Practices

Heavy metal contamination of agricultural lands as a result of various cultural practices has been documented a number of times in the literature. These practices fall into two discrete categories: 1) The use of pesticide formulations containing metals as an active ingredient, and 2) the use of contaminated sewage sludges, solid wastes or fertilizers as soil amendments. The impacts of these sources in terms of resultant contamination of soils and vegetation are discussed below.

3.2.1. Metal-containing pesticides

In some countries, metal-containing pesticides can be a major means of contamination of agricultural soils with heavy metals. In Ontario, Canada, such pesticides are the major source of soil contamination through agricultural practices, particularly in fruit-growing areas, where sprays containing such chemicals as lead or calcium arsenate or copper sulphate (Bordeaux mixture) have been frequently used to combat fungal pathogens, especially in past years (Frank *et al.*, 1976*a*,*b*). Table 2.5 lists such metal-containing pesticides that have been commonly used in Ontario, Canada, together with calculated loading rates at recommended dosages.

Various authors have documented cases of soil contamination through the use of metal-containing pesticides. In a study of heavy metal levels in soils and crops of the Holland Marsh, Ontario, Canada, Hutchinson *et al.* (1974) found that organic muck soils which had been cultivated for 40 years had elevated levels of copper in the surface 5 cm, with mean values of 135 ppm (dw), falling to 25 ppm at a depth of 22 cm. In remnant undrained parts of this extensive marsh, copper levels were <20 ppm. Surface elevations were also noted for cadmium and zinc. At this site, the metal additions occurred as copper-containing sprays used to prevent fungal infections of emerging crops. Similarly, Miles (1968) analysed soils from apple orchards in Ontario, and found arsenic levels ranging from 15·6 to 121 ppm (dw) in soils under trees, compared with <10 ppm in unsprayed areas. In another Ontario study, Frank *et al.* (1976*a*) found concentrations

TABLE 2.5

METAL-CONTAINING PESTICIDES RECOMMENDED FOR USE IN ONTARIO, CANADA FOR VARIOUS FRUIT CROPS OVER 1892–1975, AND CALCULATED ANNUAL LOADING RATES OF METALS. (AFTER FRANK *et al.* (1976*a*))

Common name of pesticide	Annual application (kg/ha-yr)	Years when recommended	Fruit crops sprayed
Inorganics			
Calcium arsenate	2·0–2·5 As	1910–1953	apple, sour cherry
Copper acetoarsenate	0·8–1·7 As	1895–1920	apple, sour cherry, peach, grape
	0·2–0·4 Cu		
Copper sulphate–calcium hydroxide	1·3–1·6 Cu	1892–1975	apple, sour cherry, peach, grape
Fixed copper[a]	1·0–3·0 Cu	1940–1975	apple, sour cherry
Lead arsenate	4·0–8·7 Pb	1910–1975	apple, sour cherry, sweet cherry, peach, grape
	1·0–2·7 As		
Zinc sulphate	5·5–7·5 Zn	1939–1955	peach
Metallic organics			
Ferbam	0·3–0·8 Fe	1948–1975	apple, sweet and sour cherry, peach, grape
Maneb, Mancozeb	0·4–0·5 Mn	1964–1975	apple, grape
Phenyl mercuric acetate	0·07–0·10 Hg	1954–1973	apple
Zineb	0·6–0·7 Zn	1957–1975	apple

[a] Various mixtures of copper chloride and/or copper sulphate with copper hydroxide.

of up to 890 ppm (dw) lead and 126 ppm (dw) arsenic in surface soils (0–15 cm) in a survey of 31 apple orchards. These metals originated with the use of lead arsenate pesticides over time periods of up to 70 years. In this study, the soil metal concentrations showed a strong relationship with orchard age (Fig. 2.1).

Mercury inputs to agricultural soils also occur, mainly from the use of organic mercurials as a seed treatment, of mercury sulphate as a root dip for

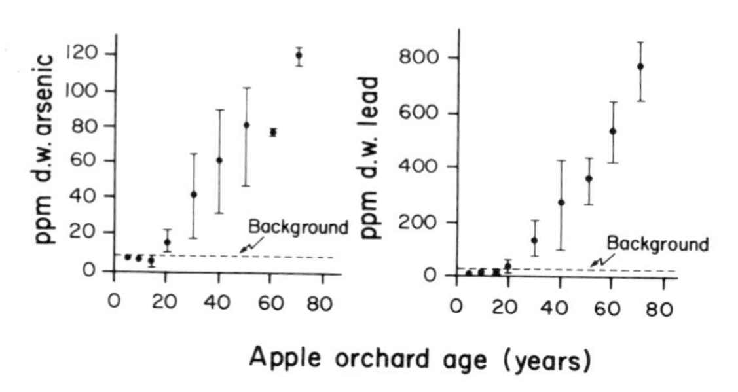

FIG. 2.1 The accumulations of arsenic and lead in the surface soils (0–15 cm) of 31 apple orchards in Ontario, Canada ranging in age from 5 to 70 years. Lead arsenate has been applied at various rates over the lines of the orchards. Mean values and ranges indicated. (After Frank *et al.* (1976*a*).)

cruciferous crops, or phenyl mercuric acetate for treatment of apple scab (Frank *et al.*, 1976*a*,*b*). Mercurial compounds have also been used for the control of turfgrass diseases. MacLean *et al.* (1973) found mercury concentrations in golf green soils (0–5 cm) ranging from 24 to 120 ppm (dw) in Ottawa, Canada.

The use of mercury compounds as agricultural seed dressings has had effects beyond those of soil contamination, with well-documented evidence of mercury accumulations and toxicity in avian and mammalian seed eaters, and avian predators of these herbivores (e.g. Tejning, 1967; Fimreite, 1970; Fimreite *et al.*, 1970; Johnels *et al.*, 1979). Of especial significance was the use of alkyl mercury compounds (such as methylmercury), since mercury present in this form is readily assimilated by animals from their foodstuffs. Fimreite *et al.* (1970), for example, compared the mercury concentrations in tissues of seed-eating rodents and birds and their avian predators with alkyl-mercury treated and untreated areas in western Canada, and found notable differences (Table 2.6).

The use of alkyl mercury compounds as seed dressings has been

TABLE 2.6

MERCURY CONTENTS OF TISSUES OF SEED-EATING RODENTS AND BIRDS AND THEIR AVIAN PREDATORS IN ALKYL-MERCURY TREATED AND UNTREATED AREAS OF WESTERN CANADA. (AFTER FIMREITE *et al.* (1970))

Organism	Mercury content (ppm (dw), $\bar{x} \pm$ S.E., n) Treated areas	Untreated areas
Rodents[a]	$1{\cdot}25 \pm 0{\cdot}68$, $n = 6$	$0{\cdot}18 \pm 0{\cdot}15$, $n = 5$
Songbirds[a]	$1{\cdot}63 \pm 1{\cdot}00$, $n = 10$	$0{\cdot}03 \pm 0{\cdot}01$, $n = 3$
Upland game birds[a]	$1{\cdot}88 \pm 0{\cdot}44$, $n = 19$	$0{\cdot}35 \pm 0{\cdot}22$, $n = 12$
All seed eaters[a]	$1{\cdot}70 \pm 0{\cdot}38$, $n = 35$	$0{\cdot}26 \pm 0{\cdot}14$, $n = 20$
Predatory birds[b]	$0{\cdot}24 \pm 0{\cdot}04$, $n = 89$	$0{\cdot}10 \pm 0{\cdot}02$, $n = 34$
Seed-eating prey[c]	$1{\cdot}16 \pm 0{\cdot}23$, $n = 61$	$0{\cdot}37 \pm 0{\cdot}15$, $n = 32$

[a] Mercury in liver, treated v. untreated areas in Alberta, Canada.
[b] Mercury in egg, treated areas in Alberta v. untreated areas in Saskatchewan, Canada.
[c] Mercury in liver, treated areas in Alberta v. untreated areas in Saskatchewan, Canada.

prohibited in most countries since the mid 1960s, when their ecological problems became recognized. Their use was prohibited in Sweden in 1966, and the less-toxic alkoxyl–alkyl mercury compounds were approved as replacements. This has led to significant declines in mercury contents of previously affected species, such as raptorial birds (Fig. 2.2a and b).

3.2.2. Metal-contaminated soil amendments

The use of metal-contaminated sewage sludges, solid wastes or fertilizers as soil amendments may also cause significant contamination of agricultural soils and crops under some conditions.

Sewage sludges are by-products of the secondary treatment of municipal sewage. Because of their high contents of plant macronutrients (nitrogen, phosphorus and potassium) and their favourable soil-conditioning properties (due to a high organic matter content) they are frequently disposed of by application to agricultural lands. For example, of $19{\cdot}5 \times 10^6$ litres (2–9 % solids) of sewage sludge liquids produced per day in Ontario, Canada in 1974, some 41 % were disposed of by application to agricultural lands (Black and Schmidtke, 1974). This high rate of usage of sewage sludges is probably typical of most industrialized countries possessing wastewater treatment facilities.

Unfortunately, many of these sludges, particularly those derived from sewage having significant industrial inputs, contain considerable quantities

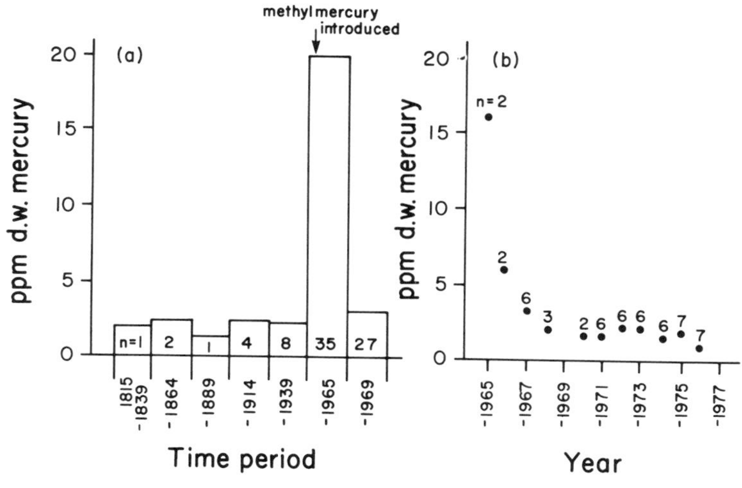

FIG. 2.2 (a) Mercury content of feathers taken from female Goshawk (*Accipiter gentilis*) collected at nests in Sweden during April–June over various time periods. (b) Mercury content of tail feathers of young Marsh Harriers (*Circus aeruginosus*) from Kuismaren, central Sweden. (After Johnels *et al.* (1979).)

of heavy metals. Table 2.7 summarizes metal analyses for a wide variety of sewage sludges from several industrialized countries. These data indicate that metal concentrations in sludges from different locations can vary tremendously, and that concentrations in some sludges reach very high values. These observations likely reflect differences in the nature of the industrial inputs to the particular wastewater systems. The metals in sewage sludge that are most likely to cause phytotoxicity problems when applied to agricultural lands are zinc, nickel, copper and cadmium (Page, 1974; Keeney, 1975).

Various proportions of the metals present in sewage sludges are available for plant uptake. This process depends on such variables as soil cation-exchange capacity, soil pH, amount of sludge applied (and its metal composition), and plant species or variety. Gaynor and Halstead (1976) presented data that showed large increases in 'available' soil, heavy metals in soils amended with sewage sludges (Table 2.8), and various other authors have found similar results (e.g. Page, 1974; Lagerwerff *et al.*, 1976; MacLean and Dekker, 1978). Many authors have grown crop plants on sludge-amended soils, and most report elevated metal concentrations in tissues, and occasionally metal toxicity (Table 2.9; see also Page, 1974;

TABLE 2.7

METAL CONTENTS OF SEWAGE SLUDGES FROM A WIDE VARIETY OF LOCATIONS (ppm (dw))

Metal	Sweden[a] Mean (range)	Michigan[b] Mean (range)	UK[c] Mean (range)	North America, Europe[d] (Range)	Ontario[e] Mean (range)
Ag	—	—	32 (5–150)	5–150	25 (4–60)
As	—	7·8 (1·6–18)	—	1–18	—
B	—	—	70 (15–1 000)	—	—
Ba	—	—	1 700 (150–4 000)	—	—
Cd	12·7 (2·3–171)	74 (2–1 100)	<200 (<60–1 500)	1–1 500	29 (2–147)
Co	15·2 (2·0–113)	—	24 (2–260)	2–260	—
Cr	872 (20–40 615)	2 031 (22–30 000)	980 (40–8 800)	20–40 000	4 200 (16–16 000)
Cu	791 (52–3 300)	1 024 (84–10 400)	970 (200–8 000)	52–11 700	1 100 (162–3 000)
Hg	6·0 (<0·1–55)	5·5 (0·1–56)	—	0·1–56	9 (1–24)
Mn	517 (73–3 861)	—	500 (150–2 500)	60–3 860	310 (60–500)
Mo	—	—	7 (2–30)	2–1 000	—
Ni	121 (16–2 120)	371 (12–2 800)	510 (20–5 300)	10–53 000	390 (7–1 500)
Pb	281 (52–2 914)	1 380 (80–26 000)	820 (120–3 000)	15–26 000	1 200 (85–4 000)
Sn	—	—	160 (40–700)	—	—
V	—	—	75 (20–400)	40–700	—
Zn	2 055 (705–14 700)	3 315 (72–16 400)	4 100 (700–49 000)	72–49 000	4 500 (610–19 000)

[a] 93 plants, after Berggren and Oden (1972).
[b] 57 plants, after Blakeslee (1973).
[c] 42 plants, after Berrow and Webber (1972).
[d] 300 plants, after Page (1974).
[e] 10 plants, after Van Loon (1974).

TABLE 2.8

INFLUENCE OF SLUDGE ADDITION ON AVAILABLE METAL CONCENTRATIONS IN THREE SOIL TYPES (ppm (dw), EXTRACTED WITH 0·005 M DTPA). (AFTER GAYNOR AND HALSTEAD (1976))

Soil type	Treatment[a]	Zn	Cu	Pb	Cd
Sandy loam (a)	control	7·7	4·4	3·4	0·37
	+ sludge	52·8	11·5	6·2	0·70
Sandy loam (b)	control	1·0	0·7	1·6	0·07
	+ sludge	31·1	4·9	3·5	0·25
Clay	control	5·7	2·2	1·8	0·14
	+ sludge	74·6	12·8	5·5	0·72

[a] 3·3 kg (dw) sludge added to 82 kg soil. Eight week incubation. Metal content of sludge: Cd 30 ppm, Pb 200 ppm, Ni 360 ppm, Cu 539 ppm, Zn 3200 ppm.

Cunningham *et al.*, 1975*a*,*b*; Webber and Beauchamp, 1975; Sidle *et al.*, 1976; Cohen *et al.*, 1978; MacLean and Dekker, 1978). These assimilated metals can then enter the human food chain through direct ingestion of produce, or indirectly if the crops are first fed to livestock. To regulate this potential problem, various government agencies in industrialized countries have proposed guidelines as to the maximum amounts of particular metals which can be added to agricultural soils with sewage sludges. Table 2.10 summarizes the proposed US Environmental

TABLE 2.9

EFFECTS OF ADDITIONS OF SEWAGE SLUDGES AT VARIOUS RATES ON THE METAL CONTENTS OF CORN AND RYE PLANTS. AVERAGE OF FOUR DIFFERENT SEWAGE SLUDGES AND TWO EXPERIMENTS USING CORN, AND ONE USING RYE. (AFTER CUNNINGHAM *et al.* (1975*a*))

Application rate (MT/ha)	Concentration in plant (ppm (dw))					
	Cd	Cr	Cu	Mn	Ni	Zn
Control	0·4	<3·0	7·4	33·3	<4·5	38
63	1·6	<3·0	14·4	44·1	<4·5	149
125	3·7	<3·0	15·8	109·9	<4·5	206
251	4·1	3·2	19·1	376·2	6·5	251
502	4·8	5·5	23·3	345·7	16·3	289
Mean metal concentration in sludges (ppm (dw))	194	9 200	1 200	600	860	5 100

TABLE 2.10

PROPOSED REGULATIONS OF UNITED STATES ENVIRONMENTAL PROTECTION AGENCY (1977) FOR ADDITIONS OF SEWAGE SLUDGES TO AGRICULTURAL SOILS. METAL DATA (kg/ha) ARE MAXIMUM AMOUNTS TO BE CUMULATIVELY APPLIED TO SOILS OF VARIOUS CATION-EXCHANGE CAPACITIES. (AFTER ZENZ *et al.* (1978))

Metal	Cation-exchange capacity (meq/100 g)		
	0–5	5–15	>15
Cd	5	10	20
Cu	125	250	500
Pb	500	1 000	2 000
Ni	50	100	200
Zn	250	500	1 000

Protection Agency regulations for five metals. These regulations take into account the ability of soils having high cation-exchange capacities to complex larger amounts of metals in a relatively unavailable form, i.e. high cation-exchange capacity soils can accommodate higher rates of metal loading without becoming toxic. Table 2.11 presents similar proposed

TABLE 2.11

PROPOSED ONTARIO, CANADA RECOMMENDATIONS FOR MAXIMUM HEAVY METAL CONCENTRATIONS IN AGRICULTURAL SOILS, AND MAXIMUM METAL ADDITIONS IN SEWAGE SLUDGES OR OTHER SOIL AMENDMENTS. (AFTER SETO AND DEANGELIS (1978))

Metal	Mean concentration in Ontario soils (ppm (dw))	Max. recommended conc. in soil (ppm (dw))	Max. recommended addition to soil (kg/ha)
As	6·5	13	15
Cd	0·7	1·4	1·6
Co	4·5	18	30
Cr	14·0	112	220
Cu	25·0	100	168
Hg	0·08	0·5	0·9
Mo	0·4	1·6	2·7
Ni	16·0	32	36
Pb	14·0	56	94
Se	0·4	1·6	2·7
Zn	54·0	216	363

guidelines for Ontario, Canada, together with proposed guidelines for maximum metal concentrations in agricultural soils. It is interesting to consider the data of these two tables in relation to Table 2.12, which presents calculated loading rates of various heavy metals under several rates of application of an 'average' sewage sludge. It is readily apparent that, for many metals, the proposed guidelines presented in Tables 2.10 and 2.11 are quickly exceeded, frequently in less than five years. Obviously, the

TABLE 2.12

AMOUNTS OF METALS ADDED TO SOILS TO A DEPTH OF 15 cm FROM VARIOUS APPLICATIONS OF A TYPICAL DOMESTIC SEWAGE SLUDGE. (AFTER PAGE (1974))

Metal	Concentration in sludge (ppm)	Amount applied to soil (kg/ha)		
		50 MT/ha	100 MT/ha	250 MT/ha
Ag	10	0·5	1	2·5
As	5	0·25	0·5	1·25
B	50	2·5	5	12·5
Ba	1 000	50	100	250
Cd	10	0·5	1	2·5
Co	10	0·5	1	2·5
Cr	200	10	20	50
Cu	500	25	50	125
Hg	5	0·25	0·5	1·25
Mn	500	25	50	125
Mo	5	0·25	0·5	1·25
Ni	50	2·5	5	12·5
Pb	500	25	50	125
Se	1	0·05	0·1	0·25
Sn	100	5	10	25
V	50	2·5	5	12·5
Zn	2 000	100	200	500

amounts of metals applied to agricultural lands in sewage sludges, and the metal concentrations in crops subsequently grown on these lands, will require close monitoring in the future (see also Chapter 5).

Other frequently used agricultural amendments also contain heavy metals, but in general their concentrations are much less than those found in sewage sludges, and their contribution to the metal burdens of agricultural soils are less.

Metal analyses for some commonly used agricultural fertilizers are presented in Table 2.13. In general, the metal contents of these formulated

TABLE 2.13

HEAVY METAL CONTENTS OF A VARIETY OF SOIL FERTILIZERS AND CONDITIONERS (ppm (dw))

Material	Cd	Co	Cr	Cu	Fe	Ni	Pb	Zn
[a]Monoammonium phosphate (12·5–50–0)	5·7	4	66	2·9	11 800	39	<3	69
[a]Diammonium phosphate (18–46–0)	5·6	4	68	2·6	11 200	37	<3	71
[a]Superphosphate (0–20–0)	2·1	4	39	2·4	5 700	23	<3	42
[a]Triple super-phosphate (0–46–0)	9·3	5	92	3·1	10 800	36	3	108
[a]Urea (46–0–0)	<0·1	<1	<3	<0·4	<3	<1	<3	<1
[a]Ammonium nitrate (34–0–0)	<0·2	<1	<5	<0·3	180	7	<3	<3
[a]Potash (0–0–60)	<0·1	2	<3	<0·6	360	4	3	<1
[a]Potassium sulphate (0–0–50)	<0·2	1	<3	<0·5	540	5	<3	<3
[a]Agricultural dolomite	<0·1	<1	<3	<0·2	120	5	<3	<2
[b]Cow manure	0·8	5·9	56	62	15 900	29	16	71

[a] After Whitby *et al.* (1977).
[b] After Furr *et al.* (1976).

fertilizers are relatively low, and their application to agricultural lands results in low metal additions. Whitby *et al.* (1977) presented data on metal loading rates for six agricultural watersheds in southern Ontario, Canada, and found much lower metal inputs from the use of fertilizers than from atmospheric loading, or from sludge or manure applications (Table 2.14).

3.3. Mining Activities

The discarding of overburden, mine excavations or other waste material from mining activities has led to severe metal contamination of many scattered, but usually relatively localized areas. Of especial significance are metalliferous mine spoils, metalliferous tailings and coal mine spoils. These three are dealt with separately below.

TABLE 2.14

METAL LOADING TO AGRICULTURAL WATERSHEDS IN SOUTHWESTERN ONTARIO, CANADA. MEAN VALUES OF SIX WATERSHEDS. DATA ARE g/ha-yr. (AFTER WHITBY *et al.* (1977))

Input	Cd	Cu	Ni	Pb	Zn
Wet + dry atmospheric inputs	—	86	—	87	532
Fertilizer applications	0·39	0·13	1·43	<0·1	3·26
Sludge and manure applications	15	461	83	275	3 460

3.3.1. Metalliferous mine spoils

Soil pollution at metal mine sites, where metal-rich overburden, excavation wastes, etc., have been dumped, is a well-documented phenomenon. A major reason for this is that such sites provide an excellent natural laboratory where the results of selection for metal tolerance can be studied. The high metal concentrations in these soils (see Table 2.15 for examples) act as powerful agents of natural selection, selecting out genetically adapted metal-tolerant ecotypes of various plant species. These are able to persist in these metal-stressed environments, whereas non-tolerant populations of the same species are quickly eliminated. Because metal concentrations on these sites are often very high (and thus selective forces are strong), evolution can occur rapidly. In addition, because the metal contaminations are often sharply discontinuous between tips and uncontaminated soils, the metal-tolerant ecotypes may in some cases be found immediately adjacent to non-tolerant ecotypes of the same species. This affords an excellent opportunity to study the process of gene flow. Some of the metal-tolerant ecotypes of various plant species are of more than evolutionary interest, as they may be used in large-scale revegetation schemes of other metal-contaminated sites, for which they are pre-adapted (providing that the contaminating metals are similar). The literature on the subject of metal-tolerant plant ecotypes is voluminous, and will not be reviewed here. Several key references are: Antonovics *et al.* (1971), Bradshaw (1975), Ernst (1975), Humphreys and Bradshaw (1976) and Smith and Bradshaw (1979).

3.3.2. Metalliferous tailings

Tailings are a waste by-product of the industrial process called milling. In this process, raw materials (e.g. metal ores) are ground to a fine powder, and are then separated either magnetically or by chemical flotation into metal-rich fractions (which are then fed to smelters for further concentrating), and

TABLE 2.15
REPRESENTATIVE SOIL METAL ANALYSES CITED IN STUDIES INVESTIGATING METALLIFEROUS MINE WASTE SITES (ppm (dw))

Study site		Cu	Zn	Pb	Reference
Garden surface soils, near operating mine, British Columbia, Canada (mean)	Site 1:	90	1 500	500	Warren *et al.* (1969)
	Site 2:	110	390	180	
Surface soils, near abandoned mines, British Columbia, Canada (mean)	Mine 1:	680	120	2	Warren *et al.* (1969)
	Mine 2:	120	130	40	
	Mine 3:	160	390	300	
Surface soils, near abandoned mine, West Devonshire, UK (mean)		240	1 750	960	Warren *et al.* (1969)
Surface soils, abandoned mine, Wales, UK (mean)			1 270	21 300	Williams *et al.* (1977)
Surface soils, abandoned mines, Wales, UK (range)	Mine 1:			1 100–1 750	Jain and Bradshaw (1966)
	Mine 2:		4 500–5 000		
	Mine 3:	680–2 700			
Surface soils, nine abandoned mines, Wales, UK (range)		90–2 300	75–40 000	80–3 600	Gregory and Bradshaw (1965)

waste milled materials, i.e. the tailings. Because this separation is not complete, some metals usually remain with the tailings and, in the case of tailings from base metal mills, these may occur in high concentrations and make attempts at revegetating disposal sites difficult. Further toxicity problems develop if the metals occur as sulphides, since large amounts of acidity are generated when the sulphides are oxidized by microbial action. Table 2.16 lists metal analyses of tailings from a number of disposal sites in Canada and the US. All of these tailings contain very high concentrations of some metals, but the problem metals differ from site to site, depending on the nature of the ore that was fed to the mill. Metal problems on the acidic tailings are particularly severe, since most metals have relatively higher solubilities at acidic pH's, and are thus more toxic.

Much research has been done on the revegetation of metal tailings dumps since apart from the usual aesthetic and economic problems associated with large areas of derelict land they may be significant local sources of fine airborne dusts when they are dry. In general, revegetation usually involves some combination of liming to raise pH and reduce metal availability, fertilizing to alleviate nutrient deficiencies, incorporation of organic materials to improve structure and sowing with various seed mixtures. Occasionally, relatively novel amendments are required, such as the use of acid or metal-tolerant plant ecotypes, inoculation with tolerant mycorrhizal fungi or the overlaying of the entire tailings area with some sort of locally available overburden, which is then revegetated. Several recent references relevant to tailings revegetation are LeRoy and Keller (1972), Dean *et al.* (1973), Peterson and Nielson (1973), Michelutti (1974), Harris and Jurgensen (1977) and Moore and Zimmerman (1977). (See also Chapter 6.)

3.3.3. Coal mine spoils

Coal mine spoils include materials removed during the surface strip-mining of near-surface coal deposits, or waste materials brought to the surface in the process of subterranean shaft-mining. Often associated with these waste materials are large amounts of reduced sulphur, mainly in the form of iron pyrite, with lesser amounts of chalcopyrite. These are found in especially large quantities in those portions of the spoils that were in closest proximity to the coal-bearing veins. When these sulphides are brought into contact with oxygen, they are metabolized by bacteria in the genus *Thiobacillus* (i.e. *T. thiooxidans* and *T. ferrooxidans*), which derive energy from the oxidation of reduced sulphur or iron compounds, and in the process generate acidity. The chemical reactions involved in these processes

TABLE 2.16

CHEMICAL ANALYSES OF A VARIETY OF METALLIFEROUS TAILINGS (ppm (dw))

Site	pH	As	Cd	Cu	Ni	Pb	Zn	S	Reference
Arctic Gold and Silver, Yukon, Canada	1·8–2·0	5 200	18	140	13	952	2 400	1·1 %	Kuja (1979)
Cyprus Anvil, Yukon, Canada	3·0–3·5	400	15	613	14	130	9 600	44·4	Kuja (1979)
United Keno, Yukon, Canada	6·0–6·1	1 350	102	172	15	3 600	6 200	5·7	Kuja (1979)
Venus, Yukon, Canada	6·9–7·0	50 000	114	330	20	2 300	18 000	13·5	Kuja (1979)
Canada Tungsten, Yukon, Canada	7·0–7·1	nd[a]	0·3	1 420	22	12	288	6·5	Kuja (1979)
Pine Point, Yukon, Canada	7·1–7·2	7	2·6	33	21	1 130	1 060	4·0	Kuja (1979)
Whitehorse Copper, Yukon, Canada	8·9–9·1	15	0·7	1 710	21	9	178	0·1	Kuja (1979)
Nickel Rim, Ontario, Canada	3·1–3·3	nd[a]	0·4	392	290	18	291	0·9	Kuja (1979)
New Mexico, USA, Cu tailings[b]	2·2	—	—	>600	—	<5	21	—	Peterson and Nielson (1973)
Utah, USA U tailings[b]	2·9	—	—	475	—	16	103	—	Peterson and Nielson (1973)
Montana, USA Cu tailings[b]	5·2	—	—	195	—	<5	24	—	Peterson and Nielson (1973)

[a] nd: not detected.

[b] ppm in saturation water extract.

are summarized below (after Ahmad, 1974; Smith, 1974).

$$FeS_2 + 7/2O_2 + H_2O \rightarrow 2SO_4^{2-} + Fe^{2+} + 2H^+ \quad (1)$$

$$Fe^{2+} + 1/4O_2 + H^+ \rightarrow Fe^{3+} + 1/2H_2O \quad (2)$$

$$Fe^{3+} + 3H_2O \rightarrow Fe(OH)_3 + 3H^+ \quad (3)$$

overall:

$$FeS_2 + 15/4O_2 + 7/2H_2O \rightarrow 2H_2SO_4 + Fe(OH)_3$$

Most coal mine spoils are made quite acidic by this process. Although the spoil pH's at various sites can range from 3 to as high as 9, most range from 4·0 to 5·5 (Berg and Vogel, 1973; Down, 1975). At these high soil acidities, large quantities of heavy metals are made soluble, and are thus available for plant uptake. Of especial significance with respect to plant toxicity are the high soluble concentrations of aluminium, iron and manganese (Massey, 1972; Berg and Vogel, 1973; Rorison, 1973). In general, the soluble metal concentrations are inversely proportional to pH (Struthers, 1964; Massey, 1972; Massey and Barnhisel, 1972; Berg and Vogel, 1973). Massey (1972) established quantitative relationships between soil pH and the log of the concentration of water-soluble zinc, copper and nickel in coal mine spoils, and calculated correlation coefficients of −0·87, −0·92 and −0·66, respectively.

The reclamation of acid coal mine spoils usually involves liming to raise pH and to lower the soluble concentrations of metals. The problem is complex, however, as it is very difficult to predict the total acidity generating potential of the pyrites in the spoils, so that it is not uncommon to find an acid reversion of reclaimed spoils (Gemmell, 1977).

3.4. Emissions from Large Industrial Sources

Large industrial sources of metal contamination, as considered here, include such facilities as primary or secondary base or ferrous metal smelters, metal refineries and metal foundries. These frequently act as point sources of emissions, often having severe metal contamination of soils and vegetation in surrounding areas, which decline exponentially to background levels at more distant sites. These various sources are described below.

3.4.1. Iron and steel industry

Iron and steel industries can be significant emitters of metal particulates, particularly of oxides of iron, especially unreactive Fe_2O_3. For example, of the total 1972 emissions of particulates in Canada of $2{\cdot}1 \times 10^6$ MT/yr, the iron and steel industry accounted for $1{\cdot}4 \times 10^5$ MT/yr, or 6·7 % of the total (Anon, 1976*a*).

A major Canadian iron–steel complex is located at Hamilton, Ontario. Considerable amounts of pollutants are emitted from this facility. This results in high measurements of particulate pollution in the city, such as suspended particulates, coefficient of haze, or settleable solids (dustfall), all of which are routinely measured in and around the city (Anon, 1978*b*). Metal particulates, most of which are Fe_2O_3, occur in large amounts in dustfall at nine stations in or bordering the industrial complex, compared with two relatively uncontaminated urban control sites (Table 2.17). Vegetation samples collected near the industrial complex also contain high iron levels. Most of the iron can be removed by a simple washing technique, indicating that it exists as particulates deposited on the leaf surfaces (Table 2.18). Similar patterns of iron deposition near iron and steel industries were also found in other Canadian cities, especially Sydney, Nova Scotia (Kilotat *et al.*, 1970; Oldreive, 1976) and Windsor, Ontario (Munn *et al.*, 1969). In the case of Windsor, most of the particulates originated with industrial activities in nearby Detroit, Michigan.

TABLE 2.17
SETTLEABLE PARTICULATE IRON AT SELECTED MONITORING SITES IN HAMILTON, ONTARIO, CANADA. (AFTER ANON. (1978*b*))

Site	Mean dustfall 1970–77 ($g\ Fe_2O_3/m^2/30\ d$)	Range 1970–77 ($g\ Fe_2O_3/m^2/30\ d$)
A	1·27	0·66–1·56
B	2·30	1·64–3·04
C	2·28	1·54–2·91
D	1·20	0·63–1·76
E	1·06	0·70–1·38
F	0·83	0·35–1·37
G	1·36	0·78–1·84
H	1·72	0·74–2·22
I	1·30	0·88–1·68
J[a]	0·30	0·18–0·58
K[a]	0·29	0·11–0·43

[a] Relatively uncontaminated urban control sites. All other sites were within or bordering the industrial complex.

TABLE 2.18

IRON CONCENTRATIONS IN PLANT TISSUES COLLECTED IN HAMILTON, ONTARIO, CANADA NEAR A MAJOR IRON–STEEL INDUSTRIAL COMPLEX. (UNPUBLISHED DATA OF T. C. HUTCHINSON)

Species	Iron concentration (ppm (dw) mean ± S.D.) Unwashed foliage	Washed foliage
Poa pratensis	1 570 ± 670	370 ± 320
Syringa vulgaris	490 ± 270	240 ± 130
Acer saccharinum	1 350 ± 770	430 ± 130
Acer negundo	1 000 ± 320	330 ± 50
Picea abies (current foliage)	790 ± 390	410 ± 160
Picea abies (one-year-old foliage)	730 ± 490	270 ± 70

Uncontaminated foliage 140 ppm Fe (dw). (Bowen, 1966.)

In spite of these high emissions of Fe_2O_3, and the resultant high rates of deposition and contamination of soils and vegetation, it appears likely that the ecological effects of this pollution are minimal. This is due to the highly unreactive nature of Fe_2O_3, viz. it is insoluble except in strong acids, and is thus not bio-available. Thus, the major problems are the aesthetic and nuisance factors associated with high rates of dustfall.

Such is not necessarily the case with particulate emissions from primary smelters roasting iron sulphide ores. For example, soils and vegetation near a large smelter at Wawa, Ontario, Canada have been contaminated by emissions of iron and arsenic, with 2·9 % iron and 157 ppm (dw) arsenic being found in surface soils at a site 1·6 km from the smelter, compared with < 1 % iron and 28 ppm arsenic at three sites at distances ranging from 40 to 56 km. Snow meltwater collected within 10 km of this smelter in 1975 contained up to 88 ppm iron and 230 ppb arsenic, compared with < 0·1 ppm iron and < 5 ppb arsenic at control sites (McGovern and Balsillie, 1972; McGovern, 1976). Severe environmental degradation of terrestrial and freshwater aquatic ecosystems has occurred in the Wawa area. Most of this, however, is due to emissions of sulphur dioxide, as sulphide ores are roasted at this smelter. Sulphur dioxide fumigations have had toxic effects on forest trees and understory vegetation, and have acidified nearby lakes, mainly as a result of dry deposition of sulphur dioxide, and of sulphuric acid aerosol, and scrubbing of sulphates by wet deposition.

By comparison, no easily discernible degradation of plant communities has occurred near another primary iron smelter located near Capreol,

Ontario, Canada. This facility does not smelt sulphide ores, and thus emits no sulphur dioxide. Iron particulate emissions are considerable, however, and dustfall deposition rates at a site 2 km from the smelter as high as 1·9 g Fe_2O_3 (as Fe)/m^2/30 d were measured in 1977, compared with 0·03 g Fe_2O_3/m^2/30 d in remote locations (Freedman, 1978).

3.4.2. Primary base metal smelters

Many case studies have been made of contamination of terrestrial environments by atmospheric emissions from primary base metal smelters. References to many of these studies are listed in Table 2.19. Rather than reviewing this vast literature, the remainder of this section will describe a single well-documented case study—the effects of a large nickel–copper smelting complex at Sudbury, Ontario, Canada.

Smelting activity began in the Sudbury region about 1885, when ore was roasted using open roast beds and small smelters. In the roasting process, heaps of sulphide ores were laid over locally cut cordwood as a fuel. This was ignited, and the heat generated oxidized the sulphides, which provided enough additional exothermic heat to begin a self-maintained oxidation. These roast beds burned for two or more months, after which the metal concentrates were gathered for further smelting and refining. This process generated choking, phytotoxic ground level plumes of sulphur dioxide, sulphuric acid mists and heavy metal particulates which devastated plant communities, already severely disturbed by clear-cutting of forests to fuel the roast beds. The death of forests led to severe erosion of the initially thin soils from the slopes of hills, and caused the granitic and gneissic Precambrian Shield bedrock to be exposed. These rocks were quickly blackened and pitted by the acid plumes. During this phase of roast bed usage, as much as $2{\cdot}7 \times 10^5$ MT of sulphur dioxide were emitted at ground level per year, as were large quantities of heavy metal particulates (Holloway, 1917). These roasting techniques were gradually replaced by more efficient smelter facilities, which vented pollutants through tall smokestacks. By 1928, the use of roast beds was forbidden by an order of the Provincial government, and all roasting was carried out at three smelter facilities located near Sudbury; at Copper Cliff, Coniston and Falconbridge. At present, all smelting activity is centred at Copper Cliff, where pollutants are emitted through a 380 m 'superstack' (the world's tallest), and at Falconbridge, with a 93 m stack. The Coniston smelter was closed in 1972, with all production transferred to the Copper Cliff smelter (Winterhalder, 1978; Freedman and Hutchinson, 1979*a*).

At present (1977 data), the 380 m smokestack at Copper Cliff emits *c.*

TABLE 2.19

STUDIES FROM VARIOUS REGIONS WHICH DOCUMENT TERRESTRIAL HEAVY METAL CONTAMINATION RESULTING FROM EMISSION FROM PRIMARY BASE METAL SMELTERS

Location	Metal contaminants	References
Avonmouth, England	Zn, Pb, Cd	Burkitt *et al.* (1972); Little and Martin (1972, 1974)
El Paso, Texas, USA	Pb	Landrigan *et al.* (1975)
Finspång, Sweden	Cu, Zn, Cd, Ni	Ruhling and Tyler (1973)
Flin Flon, Manitoba, Canada	Cu, Zn, Cd	Van Loon and Beamish (1977); Freedman and Hutchinson (1979*a*)
Helena Valley, Montana, USA	Pb	Anon. (1972*a*)
Kellogg, Idaho, USA	Pb, Zn, Ag, Cd	Ragaini *et al.* (1977)
Kootenay, British Columbia, Canada	Pb, Zn	Warren *et al.* (1969)
Mezica, Yugoslavia	Pb	Kerin (1975)
New Lead Belt, Missouri, USA	Pb, Zn, Cu, Cd	Wixson and Arvik (1973); Watson (1976); Watson *et al.* (1976)
Palmerton, Pennsylvania, USA	Zn, Cd, Cu, Pb	Buchauer (1973); Jordan (1975); Jordan and Lechevalier (1975); Strojan (1978*a*,*b*)
Poland (site not named)	Zn, Pb	Godzik *et al.* (1979)
Silver Valley, Idaho, USA	Zn, Pb, Cd	Johnson *et al.* (1975)
Solano County, California, USA	Pb	Anon. (1972*b*)
Sudbury, Ontario, Canada	Ni, Cu, Co, As, Fe	(See text)
Superior, Arizona, USA	Cu, Zn, Pb, Cd	Wood and Nash (1976)
Swansea Valley, Wales	Zn, Pb, Cd, Ni, Cu	Goodman and Roberts (1971); Goodman *et al.* (1973); Roberts and Goodman (1974)
Thompson, Manitoba, Canada	Ni	Blauel and Hocking (1974)
Trail, British Columbia, Canada	Pb, Zn, Cd	Warren *et al.* (1971*a*,*b*); Schmitt *et al.* (1971); John *et al.* (1975; 1976)
Yellowknife, Northwest Territories, Canada	As	Anon. (1977); Hocking *et al.* (1978)
Zerjav, Yugoslavia	Pb	Djuric *et al.* (1971)

$1{\cdot}2 \times 10^6$ MT of sulphur dioxide (about 20% of total Canadian emissions of sulphur dioxide) and $1{\cdot}0 \times 10^4$ MT of particulates per year. These emissions are down on pre-superstack (Copper Cliff plus Coniston) emissions in 1970 of *c.* $2{\cdot}5 \times 10^6$ MT of sulphur dioxide, and $3{\cdot}4 \times 10^4$ MT of particulates per year. Emissions from the Falconbridge smelter were about $0{\cdot}20 \times 10^6$ MT in 1977. Following changes in process technology, these Falconbridge emissions will be reduced to a projected $0{\cdot}086 \times 10^6$ MT in 1979. The particulates emitted from the Copper Cliff smelter are mainly iron oxides, although nickel and copper are also present in large amounts. On two stack sampling dates in 1977, emissions of iron were 3·3MT/d, copper 0·9 MT/d and nickel 0·7 MT/d (personal communications from W. D. Ferguson and F. Gormley of International Nickel Company Ltd, and J. Weglo and Falconbridge Nickel Mines Ltd).

The fate and effects of the emission of pollutants from the Sudbury smelters into the surrounding environment are better documented than those of any other smelter in the world, with more than 125 published papers or reports dealing with various aspects of pollutant dispersal or environmental degradation in the Sudbury area. About 104 km^2 surrounding the Sudbury smelters have been classified as 'severely barren', with vegetation almost totally removed and soils eroded from slopes, leaving blackened, naked bedrock. This devastated area is surrounded by some 363 km^2 of disturbed terrain, classified as 'impoverished vegetation', with depauperate plant communities (Watson and Richardson, 1972). It is notable that much of the devastation that exists today was created by the intense fumigations of pollutants emitted by the early roast beds, with further influences from widespread deforestation by clear-cutting, and forest and slash fires. Significant reductions in local sulphur dioxide pollution were realized when all roast bed activity was replaced by smelters in 1928, and more recently after a massive increase in nickel production during and following World War II, when the 380 m Copper Cliff smokestack was commissioned in 1972. Nevertheless, significant inputs of emitted pollutants to sites close to the smelters still occur, although the magnitudes are lower than they were historically. Figure 2.3 illustrates bulk deposition measurements of iron, copper and nickel near the Copper Cliff smelter, relative to sites up to 60 km distant, during an interval in 1977. These data indicate that iron, copper and nickel depositions are very high close to the smelter, relative to control areas. The pattern of deposition shows an exponential fall-off with increasing distance from the smelter, indicating the point-source nature of the particulate pollution (Freedman and Hutchinson, 1979*b*). Other data relevant to air quality in the Sudbury

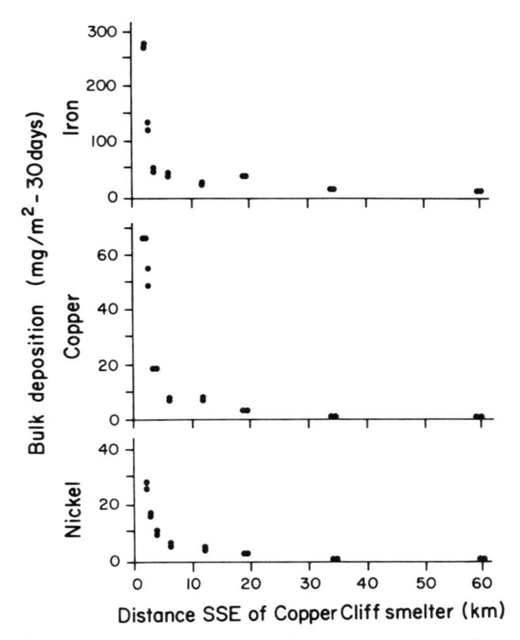

FIG. 2.3 Bulk deposition measurements from a representative sampling interval (July 18–August 16, 1977) along a SSE transect from the Copper Cliff smelter. Duplicate samples are shown. (After Freedman and Hutchinson (1979*b*).)

area are available in Costescu and Hutchinson (1972), Hutchinson and Whitby (1977), Anon (1978*c*,*d*) and Freedman and Hutchinson (1979*b*).

The continuing particulate emissions, plus metals accumulated from historical inputs, have led to severe and widespread metal contamination of soils and vegetation at sites close to the Sudbury smelters, principally by copper, nickel, cobalt and iron (Fig. 2.4, and see also Costescu and Hutchinson, 1972; Nieboer *et al.*, 1972; Hutchinson and Whitby, 1974; McGovern and Balsillie, 1972; Tomassini *et al.*, 1976; Anon, 1978*d*; Freedman and Hutchinson, 1979*b*; Rutherford and Bray, 1979). Concentrations of nickel of up to 4900 ppm (dw), and copper of up to 4900 ppm (dw) in surface litter (0–7 cm) of forested sites, and metal concentrations in vascular plants of up to 370 ppm (dw) nickel and 260 ppm copper were found at sites south of the Copper Cliff smelter in the years 1975–78 (Freedman, 1978; Freedman and Hutchinson, 1979*b*).

The high copper and nickel levels of soils close to the Sudbury smelters have been shown to be toxic to seedlings grown on them under controlled

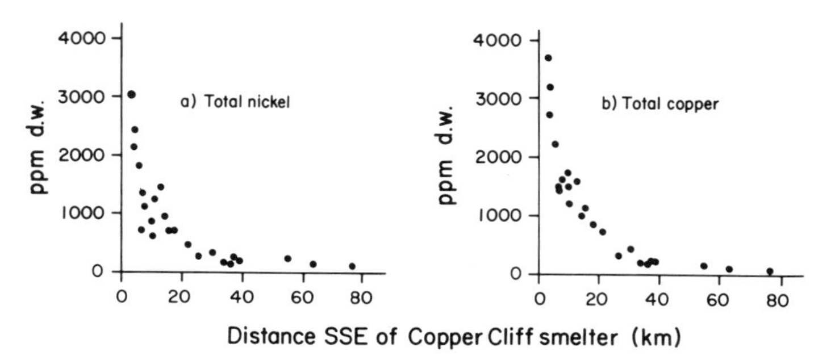

FIG. 2.4 Analyses of forest litter samples (approximately 0–7 cm) collected at forest sites at various distances from the Copper Cliff smelter. Each point represents the mean of six replicate collections. (After Freedman and Hutchinson (1979*b*).)

laboratory bioassay conditions. Whitby and Hutchinson (1974) grew a variety of plant species on metal-contaminated soils collected in the vicinity of the then-operating Coniston smelter, and found reduced growth in terms of biomass increment and seedling root elongation in all bioassay species grown on these soils compared with uncontaminated control soils. These toxic effects, which persisted in some soils even when soil acidity was neutralized by liming, were due to high concentrations of soluble copper, nickel, cobalt and aluminium in the soils. Similar symptoms were seen when the bioassay plants were grown in solutions containing pure salts of these metals. The nickel concentrations in the soils were fatal to seedling establishment of test species to a distance of 12 km from the smelter.

The high concentrations of these metals in soils have acted as a selective factor for some species. Heavy metal tolerance now occurs in several species of grasses in the Sudbury area. Hogan *et al.* (1977*a*,*b*) described copper tolerance in *Agrostis gigantea* growing on an abandoned roast bed site near Sudbury, and Cox and Hutchinson (1979) described metal co-tolerances in *Deschampsia caespitosa* populations invading the closed smelter site at Coniston, as well as at Copper Cliff. In the latter study, co-tolerance occurred for nickel, copper and aluminium, all of which were major soil contaminants at the study sites (the aluminium concentrations in the soils are high due to the high acidity, caused by sulphur emissions from the smelters). Notably, however, the Coniston populations were also tolerant to lead, cadmium and zinc, none of which occurred as soil contaminants at the Sudbury sites. Although the tolerances exhibited to the latter three metals were weaker than those for the three contaminating metals, the observations are interesting in that they suggest common physiological mechanisms of metal tolerance among different metals. This has not generally been considered the case. Rather, it has been thought that the presence of each metal at elevated concentrations in the soil was a requirement for multiple metal tolerance.

3.4.3. Secondary base metal smelters

Secondary metal smelters are industrial facilities which take discarded manufactured products and reprocess them into their constituent metals. In the process of doing so, they may emit substantial quantities of metallic particulates from smokestacks, or from fugitive sources such as ground-level piles of materials. The types of secondary smelters which have attracted the most environmental attention are those which reprocess lead products, especially automobile batteries. Table 2.20 lists references to several studies which have documented metal contamination near these industries.

TABLE 2.20

STUDIES FROM VARIOUS REGIONS WHICH DOCUMENT TERRESTRIAL HEAVY METAL CONTAMINATION RESULTING FROM PARTICULATE EMISSIONS FROM LEAD BATTERY SMELTERS

Location	Metal contaminants documented	References
Richard, British Columbia, Canada	Pb	Warren *et al.* (1971*b*)
British Columbia, Canada	Pb	John (1971)
Fraser Valley, British Columbia, Canada	Pb	John *et al.* (1972)
Toronto, Ontario, Canada	Pb, As, Sb	(See text)
Copenhagen, Denmark	Pb, Cd, Zn, Cu	Andersen *et al.* (1978)

One well-documented example of metal pollution from secondary lead smelters concerns two facilities in Toronto, Canada. An intensive research effort was expended in these studies because of the close proximities of these industries to habitations, with attendant potential health hazards.

Particulate emissions from smelter stacks and ground-level fugitive dust sources at the Toronto smelters resulted in elevated levels of suspended lead and, to a lesser extent, antimony and arsenic at sites close to the smelters (Table 2.21). Amounts of lead in settleable particulate form (i.e. dustfall) were also elevated close to the smelters, with mean values of 1780 mg lead/m^2/30 d at a site 50 m west, and 795 mg/m^2/30 d 110 m north of one of

TABLE 2.21

AMOUNTS OF SUSPENDED PARTICULATE METALS AT SITES PROXIMATE TO SECONDARY LEAD SMELTERS, BESIDE A BUSY HIGHWAY, AND IN AN URBAN CONTROL AREA. SAMPLED IN TORONTO, CANADA, SUMMER OF 1974. (AFTER PACIGA (1975))

Site	Number of samples[a]	Mean concentration (ng/m^3)		
		Lead	Arsenic	Antimony
Smelter A (1·5 m height)	25	2 310	76·7	44·5
Smelter A (rooftop)	8	11 500	89·0	59·0
Smelter B (1·5 m)	26	1 280	18·3	16·7
Expressway (1·5 m)	28	1 020	10·1	5·6
Urban control (1·5 m)	28	970	14·8	8·1

[a] 24-h Hi-volume measurement.

the smelters, compared with urban control values of 23 mg/m²/30 d. The highest dustfall measurement at the 50 m site was 3580 mg/m²/30 d. Much of the lead occurred as particles greater than 7 μm in diameter (Roberts *et al.*, 1974*a*,*b*, 1975).

The large amounts of suspended and settleable particulate lead, arsenic and antimony close to the smelters produced high concentrations of these metals in soils and vegetation (Table 2.22, and see also Roberts (1975),

TABLE 2.22

METALS IN SURFACE SOIL SAMPLES COLLECTED AT VARIOUS DISTANCES FROM TWO TORONTO, CANADA SECONDARY LEAD SMELTERS. (AFTER ROBERTS *et al.* (1974*b*))

Smelter	Distance (m)	Total metals (ppm (dw))		
		Lead	Arsenic	Antimony
Smelter A	30	42 000	1 120	1 830
	130	5 560	174	200
	230	5 160	163	180
	280	2 450	108	93
	500	610	22	12
	700	385	10	9
Smelter B	90	3 255	70	107
	110	5 845	107	257
	130	2 385	59	127
	240	900	26	30
	450	1 090	12	13
	650	190	8	8

Roberts *et al.*, 1974*a*,*b*, 1975 and Linzon *et al.*, 1976). Soil lead levels of up to 51 000 ppm (dw) were found at a site 50 m from one smelter, and arsenic and antimony levels of 1120 ppm and 1830 ppm at a 30 m site, respectively (Roberts *et al.*, 1974*b*). Uncontaminated grass turves were transferred to a household garden close to one of the smelters. The concentration of lead in the humus layer increased from an initial 145 ppm (dw), to 1100 ppm after 270 days, while grass leaves increased lead concentrations from an initial 80 ppm to 300 ppm after 25 days, remaining at this level of contamination until the experiment was terminated after 270 days (Roberts *et al.*, 1974*b*). These elevated environmental lead levels led to accumulations of lead in some of the people living in nearby residential areas. Between 13 % and 30 % of children living in the contaminated areas had absorbed excessive amounts of lead (indicated by >40 μg lead/100 ml blood and >100 ppm lead in hair), compared with <1 % in an urban control group (Roberts *et al.*, 1974*a*).

3.4.4. *Base metal refineries*

Metal refineries are facilities which take crude concentrates from smelters, and from them produce quantities of desired metals in almost pure chemical form. In the process, quantities of particulate pollutants may be emitted. Relatively few studies have been made of the environmental effects of metal refineries. One well-documented case study concerns a large nickel refinery in Port Colborne, Ontario, Canada. This facility went into production in 1922, and up to 1928 it was the site of a primary smelter as well as a nickel refinery. Emissions from the refinery are complex, i.e. there are many point-sources of emission to the atmosphere, as not all of the process waste gases are vented through a single stack. At present, all significant sources of atmospheric emissions undergo some sort of particulate removal. This was not the case in the past, however, when particulate emissions occurred at significant rates. Fugitive sources of metal contamination also exist, such as ground-level piles of materials and open railway cars, and these have also contributed substantial low-level dust emissions to the area.

These emissions have led to severe contamination of soils close to the refinery with high levels of nickel, and significant amounts of copper and cobalt (Fig. 2.5, and see also Temple, 1975, 1978*a*,*b*). Nickel concentrations in organic surface soils (0–5 cm) of up to 24 000 ppm (dw) were recorded at a site 340 m from the refinery. Contamination of vegetation has also occurred. At a site 400 m from the refinery, foliage of *Acer saccharinum* facing the refinery contained 650 ppm (dw) nickel, while foliage on the

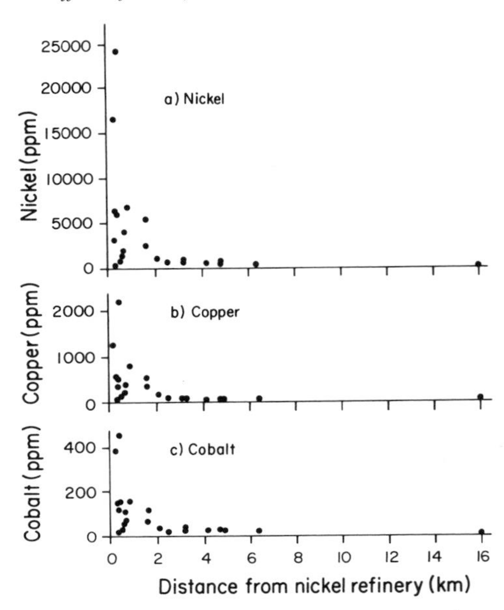

FIG. 2.5 Metals in surface soils (0–5 cm) collected at various distances from a nickel refinery at Port Colbourne, Ontario, Canada. Each point represents the mean of three replicate determinations. (After Temple (1975).)

opposite side of the same tree contained 193 ppm nickel. Similar patterns were seen with the nickel content of this species at other sites, and also with copper and cobalt. These latter patterns occurred at lower concentrations. Some foliage of *Acer saccharinum* showed symptoms of acute injury, which has been attributed to nickel toxicity (Temple, 1978*a*). A patchy distribution of a metal-related injury to cruciferous crop plants was also found at a muck farm 1 km east of the refinery. Nickel concentrations in the organic surface soils of this farm averaged 4400 ppm (dw) (eight surface 0–5 cm samples, range 1480–10 000 ppm), while copper concentrations averaged 450 ppm (164–920 ppm) and cobalt 67 ppm (25–144 ppm). These compare with surface soils at a site 16 km from the refinery having nickel concentrations of 25 ppm (dw), copper 45 ppm and cobalt 10 ppm.

3.4.5. Base metal foundries

Metal foundries are industrial complexes where metal castings are made. Few studies have been made of metal contamination around such industries. However, one series of Swedish studies has documented severe heavy metal contamination in the vicinity of a 300-year-old brass foundry, located at Gusum. Tyler (1974, 1975, 1976) analysed the surface litter in *Picea abies* forests (mor horizon) at 40 sites surrounding the foundry. At sites within 0·2–0·3 km from the source, copper concentrations were 11 000–17 000 ppm (dw), and zinc 16 000–20 000 ppm. These compare with 40–140 ppm copper and 200–600 ppm zinc at sites at about 3 km, and 12–20 ppm copper and 100–200 ppm zinc at background sites at 7–9 km. Litter metal concentrations were highly correlated with distance from the foundry, with linear regressions of log metal concentration v. distance yielding correlation coefficients of −0·934, −0·892 and −0·850 for copper, zinc and lead, respectively. These studies were primarily concerned with the effects of heavy metal contamination of forest litter horizons on the mineralization of organic matter and related microbial processes. Reductions in rates were found in most parameters measured (e.g. litter weight loss, CO_2 evolution, relative activities of various soil enzymes), and these were significantly negatively correlated with litter metal concentrations (Tyler, 1974, 1975, 1976; Ebregt and Boldewijn, 1977).

3.5. Emissions from Municipal Sources

Certain large municipal utilities, particularly coal- or oil-fired power plants and incinerators, are known to have significant emissions of metal particulates to the atmosphere. Few studies, however, have demonstrated appreciable contamination of soils or vegetation proximate to these facilities. Thus, it appears that the emitted metal particulates are well dispersed, and contribute to regional atmospheric particulate loads, particularly in urban areas. These municipal utilities are discussed below as sources of metal contamination of terrestrial environments.

3.5.1. Electric generating stations

Power generation using fossil fuels (particularly coal) to fire boilers results in large emissions of metal particulates. These particulates mainly originate from impurities present in the fuels. Tables 2.23 and 2.24 present data for the metal compositions of various coals and residual fuel oils that are used in electric generating facilities. The most significant metals in coal (in terms of total concentration) are aluminium, iron and titanium, although a wide spectrum of other metals is present, occurring in high concentrations in some samples. The most significant metals present in the residual oils used in power generation are vanadium, iron, lead and nickel.

Most generating facilities, particularly those located in urban areas, are equipped with various devices for particulate removal from flue gas

TABLE 2.23

TYPICAL METAL ANALYSES OF COALS USED FOR ELECTRIC POWER GENERATION (ppm (dw))

Element	US Coal[a]	Belgian coal[b]	Ontario coal[c]	Typical coal[d]
Ag	—	1·2	0·4	0·5
Al	—	37 600	927	10 000
As	1–16	4–25	14	5
Ba	31–380	—	137	—
Be	0·6–2·4	—	—	3
Bo	13–85	—	—	—
Cd	0·1–2·9	<5	0·10	—
Co	5–18	14	3	—
Cr	3–29	55	8	10
Cu	3–15	78	5	15
Fe	—	11 300	7 840	10 000
Hg	0·05–0·20	0·6	0·5	0·01
Li	16–78	—	—	65
Mo	1–11	2·4	—	5
Mn	10–73	60–300	24	50
Ni	4–33	55	4·1	15
Pb	4–33	—	1·9	25
Sc	—	8·9	—	5
Se	0·8–5·1	2·0	—	—
Sn	0·7–4·6	—	—	2
Ti	—	1 700	473	500
V	10–35	72	20	25
Zn	10–140	170	—	50

[a] Range of averages of samples from 13 states. (After Dvorak and Lewis (1978).)
[b] Average or range of samples from 12 mines. (After Block and Dams (1975).)
[c] Mean analysis of composite samples. (After Anon. (1976*b*).)
[d] After Bertine and Goldberg (1971).

streams. Typical mean operating efficiencies of the devices used in coal-fired power plants (i.e. % particulate mass removal) are *c*. 94%, and usually >99% under optimal operating efficiencies (Vandegrift *et al.*, 1971; Klein *et al.*, 1975*a*). In spite of these high particulate-removal efficiencies in most generating stations, large amounts of particulates are nevertheless emitted, because of the tremendous amounts of fuel that are consumed (US coal consumption in 1976 was 405×10^6 MT/yr, and is projected to reach 764×10^6 MT/yr in 1985 (Coles *et al.*, 1979)). For example, Vandegrift *et al.* (1971) calculated particulate emissions from US fossil-fuelled electric generating facilities for 1970. Coal-fired utilities emitted some

TABLE 2.24

TYPICAL METAL ANALYSES OF RESIDUAL OILS USED FOR ELECTRIC POWER GENERATION

Element	Ontario oil[a]	Typical oil[b]
Ag	—	0·001
Al	—	0·5
As	—	0·01
Cd	—	0·01
Co	—	0·2
Cu	—	0·14
Fe	66	2·5
Hg	—	10
Mo	—	10
Ni	33	10
Pb	43	0·3
Sn	—	0·01
V	280	50
Zn	—	0·25

[a] Residual fuel oil, based on Venezuela crude oil. (After Anon. (1976*b*).)
[b] After Bertine and Goldberg (1971).

$2{\cdot}8 \times 10^6$ MT/yr, while oil-fired and gas-fired facilities emitted $0{\cdot}033 \times 10^6$ and $0{\cdot}022 \times 10^6$ MT/yr, respectively. An additional $2{\cdot}4 \times 10^6$ MT/yr of particulates were emitted from coal-fired industrial boilers. Vandegrift *et al.* (1971) also presented calculations of future particulate emissions from US electrical utilities, which projected large increases from the 1970 total emissions of $3{\cdot}05 \times 10^6$ MT/yr, to some $4{\cdot}0 \times 10^6$ MT/yr in 1980, $5{\cdot}4 \times 10^6$ MT/yr in 1990, and $7{\cdot}0 \times 10^6$ MT/yr by the year 2000. Thus, total particulate emissions from electricity generation are considerable, and they are likely to increase in the future as generating capacities increase, and as coal-fired facilities become more prevalent in some areas.

These particulates contain large quantities of aluminium and iron, but a wide spectrum of other heavy metals is also present. Table 2.25 describes rates of metal emissions from coal-fired power plants from various locations. These various stations differ in generating capacity, particulate emission control technology and metal composition of fuels, and these all contribute to their very different rates of emissions of the various metals.

Significantly, most of the particulates emitted by power plants equipped with emission control facilities are very small ($<10\,\mu$m diameter), due to

TABLE 2.25

METAL EMISSIONS FROM COAL-FIRED POWER PLANTS FROM VARIOUS LOCATIONS (kg/yr). POWER PLANTS DIFFER IN METAL COMPOSITION OF FUEL, GENERATING CAPACITY AND PARTICULATE EMISSION CONTROL TECHNOLOGY

Element	US[a] 100 MW	US[b] 83 MW	US[c] 870 MW	US[d] 1 580 MW	US[e] 1 400 MW	US[f] 750 MW	US[f] 350 MW	US[f] 250 MW	Denmark[g] 275 MW
Ag	—	—	—	—	—	6	0·7	<4·5	—
Al	—	—	—	—	—	133 000	56 000	750 000	—
As	40	4 330	105	136	1 198	173	1·4	1 250	—
Ba	120	5 243	158	7 570	284	3 300	<7	8 000	—
Be	—	—	—	—	—	10	8	33	—
Bo	1 790	—	—	—	—	5 400	1 575	71 300	—
Cd	30	—	11	<9	120	65	2·8	73	15
Co	40	948	11	27	120	105	27	325	—
Cr	130	2 265	158	145	73	6 150	70	4 750	145
Cu	—	—	—	—	—	473	221	2 275	210
Fe	—	—	—	—	—	72 000	17 500	1 000 000	—
Hg	60	—	53	109	233	180	30	109	20
Mo	510	—	—	—	—	2 900	56	3 750	—
Mn	160	4 396	105	200	12 000	1 800	315	7 750	830
Ni	2 190	—	—	—	—	825	525	3 500	310
Pb	480	—	105	327	2 365	405	189	400	30
Sc	—	526	5	27	237	—	—	—	—
Se	110	—	210	927	237	143	221	375	—
Ti	—	—	—	—	—	6 450	3 850	21 500	—
V	140	3 265	210	291	2 365	3 300	455	3 250	100
Zn	100	—	1 051	400	12 000	2 000	150	3 750	520

[a] After Dvorak and Lewis (1978).
[b] After Horton and Dorsett (1976).
[c] After Klein *et al.* (1975*a*,*b*).
[d] After Turner and Strojan (1978).
[e] After Vaughan *et al.* (1975).
[f] After Schwitzgebel *et al.* (1975).
[g] After Pilegaard (1976).
See also Ondov *et al.* (1979*a*,*b*).

size selection by the control devices (Natusch *et al.*, 1975; Turner and Strojan, 1978; Coles *et al.*, 1979; Smith *et al.*, 1979). Because of this small size (which gives the particulates aerodynamic properties similar to those of gases, and thus very low deposition velocities), and the fact that most emissions are from tall smokestacks, the particulates tend to be very well dispersed from their sources.

These characteristics of the metal particulates emitted from power plants result in relatively low rates of deposition to sites proximate to the stations. Thus, the various studies which have attempted to show metal contamination of soils and vegetation close to power plants have produced ambiguous results (see Table 2.26). Even when a relationship with distance has been found, the elevations over background levels close to the sources are relatively minor, and no acute ecological problems are associated with the contamination.

Thus, the metal particulate emissions from power stations contribute to regional atmospheric contaminations rather than acting as local point-sources of pollution. They thus contribute to the generally higher metal contents of urban air, relative to rural locations (Dubois *et al.*, 1966; Van Loon, 1973; Paciga, 1975; Andersen *et al.*, 1978). The role of coal-fired power plants in contributing to regional depositions of mercury may be of especial significance (Joensuu, 1971). Although present in relatively low concentrations in coal (ranging from 0·01 to 0·6 ppm (dw) in Table 2.23; note, however, that Joensuu (1971) calculated an average mercury concentration in 36 US coals of 3·3 ppm (dw) by using analytical techniques where volatilization was less significant), most of the mercury is vaporized during combustion of the coal, and in this form is not removed by pollution control devices. Various estimates of mercury emissions from coal-fired power plants range from 87 % to 98 % of the mercury originally present in the fuel (Anderson and Smith, 1977; Klein *et al.*, 1975*a*,*b*; Schwitzgebel *et al.*, 1975). The ecological significance of the contributions of mercury emissions from coal-fired generating stations, relative to other man-effected and natural sources, is at present uncertain. This is largely due to variations in the calculated magnitudes of the various sources. For example, Bertine and Goldberg (1971) calculated that in 1967, world mercury emissions from the burning of coal (*c.* 3×10^6 kg Hg/yr) were some 86 % of natural weathering mobilizations (estimated as $3{\cdot}5 \times 10^6$ kg Hg/yr). On the other hand, Van Horn (1975) calculated that the total 1974 US emissions of mercury from coal-fired power plants were only some 9 % of all industrial releases, and only 4 % of natural mercury outgassings (estimated as some $1{\cdot}0 \times 10^6$ kg Hg/yr).

TABLE 2.26

SUMMARY OF VARIOUS STUDIES WHICH HAVE INVESTIGATED HEAVY METAL CONTAMINATION AROUND COAL-FIRED ELECTRIC GENERATING STATIONS

Location	Capacity	Effects	References
Aiken, South Carolina, USA	83 MW	No enrichment of surface 0–7·5 cm soils Possible enrichment of *Eupatorium compositifolium* by Sc	Horton and Dorsett (1976)
Memphis, Tennessee, USA	870 MW	No enrichment of surface 0–15 cm soils No enrichment of bryophytes	Bolton *et al.* (1973)
Holland, Michigan, USA	650 MW	Surface 0–2 cm soils contaminated with Ag, Cd, Co, Cr, Cu, Fe, Hg, Ni, Ti, Zn Vegetation enriched with Cd, Fe, Ni, Zn	Klein and Russel (1973)
Glen Rock, Wyoming, USA	750 MW	Possible enrichment surface 0–2·5 cm soils with Sb and As *Artemesia tridentata* enriched with Se, Sr, U, V, possibly also Co, Ti, Zn	Anderson *et al.* (1975) Connor *et al.* (1976)
Farmington, New Mexico, USA	2 085 MW	Possible enrichment with Pb in surface 0–1·2 cm soils Possible enrichment of foliage with Pb No enrichment of surface soils with Hg	Cannon and Anderson (1972) Crockett and Kinnison (1979)
Nevada, USA	1 580 MW	No clear enrichment of surface 0–7·5 cm soils No clear enrichment of *Encelia farinosa* foliage	Bradford *et al.* (1978)
Oak Ridge, Tennessee, USA	—	Hg elevated in *Dicranum scoparium*	Huckabee (1973)
Central Illinois, USA	1 200 MW	No elevated Hg in soils, freshwater lakes, lake sediments, or fish	Anderson and Smith (1973)
Copenhagen, Denmark	275 MW	Surface soils slightly enriched in Mn, V, Zn, Ni, Cu, Pb, Cr, Cd *Achillea millefolium* tissues slightly enriched with Cr, Cu, Hg, Mn, Ni, Pb, V, Zn Mosses enriched with Hg	Pilegaard (1976)

The control of particulate pollution by power plants, particularly those that are coal-fired, results in the production of large amounts of fly ash, which must be disposed of by land filling, by use as aggregate in the construction industry, or by other methods. Various studies have presented data on the elemental composition of fly ashes from power plants (Table 2.27 and see also Natusch *et al.*, 1975, Block and Dams, 1976 and Furr *et al.*, 1977). Aluminium and iron are present in particularly high concentrations, followed by titanium, chromium, copper, manganese, nickel, lead, vanadium and zinc. Various studies have examined the growth and metal composition of plants grown on soil–fly ash mixtures (e.g. Townsend and Gillham, 1973; Furr *et al.*, 1977; Elseewi *et al.*, 1978; Martens and Beahm, 1978). In general, growth decrements are not

TABLE 2.27
RANGES OF METAL CONCENTRATIONS OF FLY ASHES COLLECTED FROM VARIOUS US COAL-FIRED GENERATING STATIONS (ppm (dw))

Element	Range[a]	Range[b]
Al	59 100–143 400	—
As	2–312	10–500
B	10–600	—
Cd	0·1–3·8	10–100
Co	5–73	10–100
Cr	43–259	10–1 000
Cu	45–616	10–1 000
Fe	27 100–290 000	—
Mn	58–543	100–2 000
Mo	7–41	10–100
Ni	2–115	10–1 000
Pb	3–241	100–5 000
Sb	1–13	1–100
Se	1–17	1–50
Sn	27–334	1–10
Sr	59–3 855	—
Ti	2 760–8 310	—
V	68–442	50–5 000
Zn	14–406	10–2 000
pH	4·2–11·8	—

[a] Range in 23 US samples. (After Furr *et al.* (1977).)
[b] Range in midwestern and western US samples. After Natusch *et al.* (1975).

observed, and metal uptake data are ambiguous, partly due to the low heavy metal solubilities in the generally circumneutral to alkaline soil–fly ash mixtures. Furr *et al.* (1977) grew cabbage on soil–fly ash mixtures, and found significant correlations between the concentrations of arsenic, boron, molybdenum, selenium and strontium in cabbage tissues, and the concentrations of those metals in fly ash. Other elements showed no correlations or negative correlations. Elseewi *et al.* (1978) made similar observations. They found higher contents of strontium and molybdenum in Swiss chard and lettuce grown on fly ash-amended soils, but lower concentrations of zinc and manganese. Martens and Beahm (1978) found no significant metal uptake by *Festuca arundinacea* grown on soil–fly ash mixtures.

3.5.2. Municipal incinerators

Municipal solid waste incinerators can also be significant emitters of metal particulates. For example, the contribution of 'all refuse burning' to urban atmospheric particulate concentrations has been estimated as *c.* 9–28 % in the Boston, US region, 18 % in New York–New Jersey, 23 % in Washington D.C., 4·2 % in the State of Maryland and 1·5 % in San Francisco. A large proportion of these particulates originate with municipal incinerators (Greenberg *et al.*, 1978*a*).

These metal emissions are due to the high concentrations present in most municipal solid wastes (Table 2.28). Metal emission rates for the two incinerators for which data are presented in Table 2.28 (see also Greenberg *et al.* (1978*a*,*b*) for data relevant to the metal composition of emitted particulates) are comparable to those that are calculated for large coal-fired electric generating utilities (cf Table 2.25). This observation reflects the higher concentrations of most metals in refuse compared with coal, and also the less efficient particulate emission controls at most incinerators, compared with most urban electric utilities. We are not aware of any studies that have investigated the metal contamination of soils of vegetation close to municipal incinerators. In view of the relatively high rates of particulate emission from these urban utilities, such investigations may well be warranted.

3.6. Emissions from Automobiles

Many authors have documented the distributions of heavy metals in soils, vegetation and the atmosphere along transects originating with busy highways. All studies have found clearly defined gradients of lead

TABLE 2.28

METALS IN COMBUSTIBLE FRACTION OF US MUNICIPAL SOLID WASTE, AND RATES OF METAL PARTICULATE EMISSION FROM TWO US MUNICIPAL INCINERATORS

Element	Metals in municipal solid waste[a] (ppm, means (range))	Emissions[b] (kg/day, means ± S.D.)	Emissions[c] (kg/day, means ± S.D.)
Ag	3 (<3–7)	0·14 ± 0·014	—
Al	9 000 (5 400–12 000)	8·6 ± 4·3	—
Ba	170 (47–450)	0·43 ± 0·29	—
Cd	9 (2–22)	0·57 ± 0·29	1·4 ± 3·9
Co	3 (<3–5)	0·007 ± 0·004	—
Cr	55 (20–100)	0·29 ± 0·14	—
Cu	350 (80–900)	1·1 ± 0·7	1·1 ± 3·1
Fe	2 300 (1 000–3 500)	4·9 ± 1·9	—
Hg	1·2 (0·66–1·9)	~0·14	—
Mn	130 (50–240)	0·86 ± 0·29	—
Ni	22 (9–90)	0·14 ± 0·04	—
Pb	330 (110–1 500)	53 ± 14	2·9 ± 7·9
Sb	45 (20–80)	1·3 ± 1·3	—
Sn	20 (<20–40)	5·7 ± 0·9	—
Zn	780 (200–2 500)	66 ± 33	8·2 ± 22·6

[a] Average and range for 5 US incinerators. (After Law and Gordon (1979).)
[b] Capacity of 270 MT solid waste/d. (After Law and Gordon (1979).)
[c] Capacity of 204 MT solid waste/d. (After Jacko and Neuendorf (1977).)

contamination beside roadways, and some have also shown generally less-well-defined gradients of other metals, including cadmium, chromium, copper, nickel, vanadium and zinc (Table 2.29, Fig. 2.6).

The source of lead contamination has been the use of leaded gasoline. Since 1923—and especially since 1946—tetraethyl lead has been added to gasoline at a rate of about 0·8 g lead/litre in order to increase engine efficiency and gasoline economy, while decreasing engine wear due to engine 'knock' (Atkins, 1969). Most of the lead in gasoline is emitted through the automobile tailpipe, at a rate of *c.* 0·07 g/km (Cantwell *et al.*, 1972), accounting for some 40% of emitted automobile particulates (Atkins, 1969). Anon (1972*c*) estimated that 98% of US atmospheric lead emissions, some $1{\cdot}6 \times 10^5$ MT lead/yr, originated with automobile exhausts. Some 39% of the emitted lead particulates have diameters of less than 1 μm and are well dispersed. The remainder settles and deposits near the source, usually within about 50 m of a roadway (Cantwell *et al.*, 1972).

The degree of urban lead contamination is presently decreasing in many

TABLE 2.29

SUMMARY OF VARIOUS STUDIES WHICH HAVE DOCUMENTED CONTAMINATION OF ROADSIDE ENVIRONMENTS BY HEAVY METALS

References	Contamination documented
Warren *et al.* (1966)	Roadside soils contaminated with lead
Ruhling and Tyler (1968)	Roadside soils contaminated with lead
Daines *et al.* (1970) Motto *et al.* (1970)	Roadside soils and vegetation contaminated with lead
Lagerwerff and Specht (1970)	Roadside soils and vegetation contaminated with lead and, to a much lesser degree, cadmium, nickel and zinc
Creason *et al.* (1971)	High concentrations of atmospheric lead particulates and high rates of lead dustfall beside busy roadways
Page *et al.* (1971)	Lead contamination of roadside atmosphere, soils and vegetation near major highways
Hutchinson (1972)	Roadside soils and vegetation contaminated with lead and, to a much lesser extent, cadmium, nickel and vanadium
Oliver and Kinrade (1972)	Soil contamination with lead at dump site for snow cleared from roads
Manson and de Koning (1973)	High concentrations of atmospheric lead particulates beside busy roads
Ward *et al.* (1977)	Roadside soils and vegetation contaminated by lead and, to a much lesser extent, cadmium, chromium, copper, nickel and zinc
Hoggan *et al.* (1978)	Contamination of urban atmosphere by lead particulates; 50 % reduction noted over 1971–76 due to widespread use of unleaded gasoline
Wheeler and Rolfe (1979)	Roadside soils and vegetation contaminated by lead

regions due to the widespread and increasing use of catalytic converters to control emissions of other automobile pollutants. Automobiles equipped with these devices require the use of unleaded gasoline, as the converter catalysts are 'poisoned' by lead. Hoggan *et al.* (1978) reported a 50 % decrease in atmospheric lead particulates in the Los Angeles, USA, area over the period 1971–76. They attributed this phenomenon to the increasing use of unleaded gasoline over that period.

The sources of the other metals which have been documented as roadside contaminants are less well defined than that of lead. They presumably

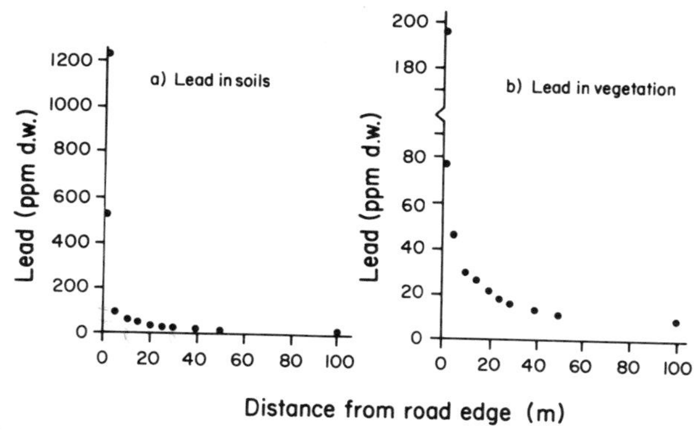

FIG. 2.6 Lead concentrations in soil and vegetation along a transect originating beside a highway with an average traffic load of 8100 vehicles/d. (After Wheeler and Rolfe (1979).)

derive from the wear of metallic automobile parts containing these metals, from the wear of tyres, and from the use of certain of these metals (e.g. nickel) as gasoline or oil additives (Lagerwerff and Specht, 1970; Hutchinson, 1972).

3.7. Other Sources

Various other sources can contribute to localized metal contaminations of the terrestrial environment, and several of these are discussed below.

Several authors have reported elevated soil metal concentrations at sites that have served as dumps for municipal solid wastes. For example, Purves (1972) found elevated soil concentrations of copper, lead, nickel and zinc at two sites in the UK and one in Italy that had received municipal solid wastes, and Warren *et al.* (1969) made similar observations for copper, lead and zinc in a reclaimed solid waste disposal site that was being used for vegetable gardens. Certainly the use of municipal refuse-derived composts as an agricultural amendment, or the agricultural use of reclaimed disposal sites, must be accompanied by a monitoring of metal levels in edible plant tissues.

Various industrial sources other than those already mentioned have been documented as causing localized incidences of metal pollution in their environs. Klein (1972) documented heavy metal contamination of soils collected within industrial districts of Grand Rapids, Michigan, USA compared with urban non-industrial or nearby agricultural areas. A general contamination was found for the metals silver, cadmium, cobalt, chromium, copper, iron, nickel, lead, zinc and mercury. These metals (with the exception of lead) were also found in elevated concentrations in soils collected near an airport. Klein (1972) also found that a local asphalt plant was serving as a point-source of mercury emissions. In another study, Page and Bingham (1973) present data which indicate soil contamination by cadmium near electroplating industries in Southern California, USA.

In another interesting case, high-voltage electricity transmission lines were linked to incidences of copper poisoning of sheep (a species which is hypersensitive to copper intoxication) grazing on forage underneath the lines (Hemkes and Hartmans, 1974). Apparently, copper leached from the transmission lines causing slight increases in the copper content of soils and forage, which posed a health hazard to sheep grazing within 20 m of both sides of the transmission corridor.

4. SUMMARY

Numerous case studies have been described of contamination of terrestrial environments by heavy metals. These include naturally occurring sources, or, more frequently, situations where man-effected emissions of pollutants have contaminated local environments. The naturally occurring sources, particularly surface metalliferous mineralizations, frequently contain metal concentrations in the thousands of ppm or higher in soils and vegetation, and thus they are comparable in degree of contamination to the worst of the man-effected situations.

Of especial significance among the various man-effected sources of contamination are emissions of various heavy metals from industrial sources, particularly from metal smelters and refineries. Local contamination of soils and vegetation around many of these industries, often in conjunction with pollution by gaseous sulphur dioxide, contributes to the degradation of surrounding natural and agronomic ecosystems. Also of significance are: (1) metal emissions from automobiles, (2) metal contamination at disposal sites of metal mine wastes, tailings and sulphide-containing coal mine spoils, and (3) agricultural practices such as the use of metal-contaminated sewage sludges as soil amendments, or the use of metal-based pesticides.

In many of the case studies described, much of the presently existing contamination has resulted from the persistence of metal pollutants emitted historically, when particulate emissions control technology was less sophisticated than at present, or was not used effectively by industry.

Present trends towards (1) effective particulate control at industrial sources by the use of such devices as electrostatic precipitators, scrubbers, or baghouses, usually with removal efficiencies of >95%, (2) the use of unleaded gasoline, and (3) declining usage of metal-based pesticides in agriculture, should significantly reduce current environmental contamination by heavy metals in many countries. Problems will remain, however, in already contaminated environments, because of the persistence of heavy metals in most terrestrial ecosystems.

The problems of the large-scale, but well dispersed, metal emissions from utilities such as coal-fired power plants or municipal incinerators (especially of volatile elements such as mercury) present special problems, the ecological significance of which at present is a long way from being understood.

REFERENCES

Ahmad, M. U. (1974). Coal mining and its effect on water quality. In: *Extraction of Minerals and Energy: Today's Dilemmas.* (R. A. Deju (ed)), Ann Arbor Science, Ann Arbor, Michigan, pp. 49–56.

Allan, R. J. (1971). Lake sediments: a medium for regional geochemical exploration of the Canadian Shield. *Can. Mining and Metall. Bull.* 43–59.

Allaway, W. H. (1968). Agronomic controls over environmental cycling of trace elements. *Adv. Agron.* **20**: 235–74.

Alstrup, V. and E. S. Hamen. (1977). Three species of lichen tolerant of high concentrations of copper. *Oikos.* **29:** 290–3.

Andersen, A., M. F. Hovmand and I. Johnsen (1978). Atmospheric heavy metal deposition in the Copenhagen area. *Environ. Pollut.* **17**: 133–51.

Anderson, B. M., J. R. Keith and J. J. Connor (1975). Antimony, arsenic, germanium, lithium, mercury, selenium, tin, and zinc in soils of the Powder River Basin. In: *Geochemical Survey of the Western Coal Regions. Annual Progress Report.* Open File Report 75–436. US Geological Survey, Denver, Colorado, pp. 50–2.

Anderson, W. L. and K. E. Smith (1977). Dynamics of mercury at coal-fired power plant and adjacent lake. *Environ. Sci. Technol.* **11**: 75–80.

Anon. (1972*a*). Helena Valley, Montana—area environmental pollution study. Office of Air Programs, US Environmental Protection Agency, Pub. AP-91.

Anon. (1972*b*). A joint study of lead contamination relative to horse deaths in southern Solano County. Air Resources Board, State of California.

Anon. (1972*c*). Lead—airborne lead in perspective. Committee on Biological Effects of Atmospheric Pollutants, National Academy of Science, Washington, D.C.

Anon. (1976*a*). A nationwide inventory of air pollutant emissions. Summary of emissions for 1972. Air Pollution Control Directorate, Environment Canada, Ottawa, Pub. EPS 3-AP-75-5. 13 pp.

Anon. (1976*a*). Atmospheric emissions from fossil-fuelled generating stations. Design and Development Division, Ontario Hydro., Toronto. 125 pp.

Anon. (1977). Task Force on Arsenic. Final Report. Yellowknife, N.W.T. Canadian Public Health Association, Ottawa. 144 pp.

Anon. (1978*a*). A nationwide inventory of emissions of air contaminants (1974). Air Pollution Control Directorate, Environment Canada, Ottawa, Pub. EPS 3-AP-78-2. 145 pp.

Anon. (1978*b*). Air quality, Hamilton, 1970–1977. Technical Support Section, West Central Region, Ontario Ministry of the Environment, Stony Creek. 44 pp.

Anon. (1978*c*). Air quality assessment studies in the Sudbury area. Vol. 1. Ambient air quality, 1976–1977. Technical Support Section, Ontario Ministry of the Environment, Northeastern Region, Sudbury. 104 pp.

Anon. (1978*d*). Air quality assessment studies in the Sudbury area. Vol. 2. Effects of sulfur dioxide and heavy metals on vegetation and soils, 1970–1977. Technical Support Section, Ontario Ministry of the Environment, Northeastern Region, Sudbury. 105 pp.

Antonovics, J., A. O. Bradshaw and R. G. Turner (1971). Heavy metal tolerance in plants. *Adv. Ecol. Res.* **7**: 1–85.

Atkins, R. P. (1969). Lead in a suburban environment. *J. Air Poll. Control Assoc.* **19**: 591–5.

Aubert, H. and M. Pinta (1977). *Trace Elements in Soils*. Elsevier Scientific Publishing Co., New York. 395 pp.

Aumento, F. (1970). Serpentine mineralogy of ultrabasic intrusions in Canada, and on the Mid-Atlantic Ridge. Geological Survey of Canada, Department of Energy, Mines and Resources, Paper 69-53, Ottawa. 51 pp.

Berg, W. A. and W. G. Vogel (1973). Toxicity of acid coal-mine spoils to plants. In: *Ecology and Reclamation of Devastated Land*, Vol. 1 (R. J. Hutnik and G. Davis (eds)), Gordon and Breach, New York, pp. 57–67.

Berggren, B. and S. Oden (1972). 1. Analysresultat Rorande Tungmetaller Och Klorerade Kolvaten I. Rotslam Fran Svenska Reningsverk 1968–1971. Institutionen fur Markvetenskap Lantbrukshogskolan, 750 07 Uppsala, Sweden.

Berrow, M. L. and J. Webber (1972). Trace elements in sewage sludges. *J. Sci. Fd. Agric.* **23**: 93–100.

Bertine, K. K. and E. D. Goldberg (1971). Fossil fuel combustion and the major sedimentary cycle. *Science*. **173**: 233–5.

Black, S. A. and N. A. Schmidtke (1974). Overview of Canadian sludge handling and land disposal practices and research. Presented at Canada/Ontario Sludge Handling and Disposal Seminar, Toronto. Sept. 18–19, 1974. 11 pp.

Blakeslee, P. A. (1973). Monitoring considerations for municipal wastewater effluent and sludge application to the land. US Environmental Protection Agency, US Dept. of Agriculture, Universities Workshop. Champaign, Urbana, Illinois. July 9–13, 1973.

Blauel, R. A. and D. Hocking (1974). Air pollution and forest decline near a nickel smelter. Northern Forest Research Centre, Environment Canada, Edmonton. Information report NOR-X-115. 39 pp.

Block, C. and R. Dams (1975). Inorganic composition of Belgian coals and coal ashes. *Environ. Sci. Technol.* **9**: 146–150.

Block, C. and R. Dams (1976). Study of fly ash emission during combustion of coal. *Environ. Sci. Technol.* **10**: 1011–17.

Bolton, N. E. *et al.* (1973). Trace element measurements at the coal-fired Allen Steam Plant. Progress Report, June 1971–January 1973. USAEC Report ORNL-NSF-EP-43. Oak Ridge National Laboratory, NTIS. Cited in Turner and Strojan (1978).

Bowen, H. J. M. (1966). *Trace Elements in Biochemistry*. Academic Press, New York. 241 pp.

Boyle, R. W. (1971). Environmental control. The land environment. *Can. Mining and Metall. Bull.*, 46–50.

Bradford, G. R., A. L. Page, I. R. Straughan and H. T. Phung (1978). A study of the deposition of fly ash on desert soils and vegetation adjacent to a coal-fired generating station. In: *Environmental Chemistry and Cycling Processes* (D. C. Adriano and I. L. Brisbin, Jr (eds)), Technical Information Center, US Dept. of Energy, Washington, pp. 383–93.

Bradshaw, A. D. (1975). The evolution of metal tolerance and its significance for vegetation establishment on metal contaminated sites. In: *Symposium Proceedings, Vol II, Part 2—International Conference on Heavy Metals in the Environment* (T. C. Hutchinson (ed)), Toronto, Canada, pp. 599–622.

Brooks, R. R., E. D. Wither and B. Zepernick (1977). Cobalt and nickel in *Rinorea* species. *Plant Soil*, **47**: 707–12.

Buchaver, M. J. (1973). Contamination of soil and vegetation near a zinc smelter by zinc, cadmium and lead. *Environ. Sci. Technol.* **7**: 131–5.

Burkitt, A., P. Lester and G. Nickless (1972). Distribution of heavy metals in the vicinity of an industrial complex. *Nature*. **238**: 327–8.

Cannon, H. L. (1960). Botanical prospecting for ore deposits. *Science*. **132**: 591–8.

Cannon, H. L. and B. M. Anderson (1972). Trace element content of the soils and vegetation in the vicinity of the Four Corners Power Plant. Part III. In: *Southwest Energy Study*. Coal Resources Work Group, US Dept. of the Interior. Cited in Turner and Strojan (1978).

Cantwell, E. N., E. S. Jacobs, W. G. Gunz and V. E. Liberi (1972). Control of particulate lead emissions from automobiles. In: *Cycling and Control of Metals* (M G. Curry and G. M. Gigliotti (eds)), US Environmental Protection Agency, Cincinnati, pp. 95–107.

Cohen, D. B., M. D. Webber and D. N. Bryant (1978). Land application of chemical sewage sludge—lysimeter studies. In: *Sludge Utilization and Disposal Conference. Proceedings No. 6*. Environment Canada, Ottawa, pp. 108–37.

Coles, D. G., R. C. Ragaini, J. M. Ondov, G. L. Fisher, D. Silberman and B. A. Prentice (1979). Chemical studies of stack fly ash from a coal-fired power plant. *Environ. Sci. Technol.* **13**: 455–9.

Connor, J. J., J. R. Keith and B. M. Anderson (1976). Trace metal variation in soils and sagebrush in the Powder River Basin, Wyoming and Montana. *J. Res. U.S. Geol. Surv.* **4**: 49–59.

Costescu, L. M. and T. C. Hutchinson (1972). The ecological consequences of soil pollution by metallic dust from the Sudbury smelters. In: *Proceedings: Environmental Progress in Science and Education*, Institute of Environmental Sciences, New York, pp. 540–3.

Cox, R. M. and T. C. Hutchinson (1979). Metal co-tolerances in the grass *Deschampsia cespitosa*. *Nature*. **279**: 231–3.

Creason, J. P., O. McNulty, L. T. Heiderscheit, D. H. Swanson and R. W. Buechley (1971). Roadside gradients in atmospheric concentrations of cadmium, lead, and zinc. In: *Trace Substances in Environmental Health—V* (D. D. Hemphill (ed)), University of Missouri, Columbia, pp. 129–42.

Crockett, A. B. and R. R. Kinnison (1979). Mercury residues in soil around a large coal-fired power plant. *Environ. Sci. Technol.* **13**: 712–15.

Cunningham, J. D., D. R. Keeney and J. A. Ryan (1975*a*). Yield and metal composition of corn and rye grown on sewage sludge-amended soil. *J. Environ. Qual.* **4**: 448–54.

Cunningham, J. D., J. A. Ryan and D. R. Keeney (1975*b*). Phytotoxicity in and metal uptake from soil treated with metal-amended sewage sludge. *J. Environ. Qual.* **4**: 455–60.

Daines, R., H. Motto and D. M. Chilko (1970). Atmospheric lead: Its relationship to traffic volume and proximity to highways. *Environ. Sci. Technol.* **4**: 318–22.

Dean, L. C., R. Havens, K. T. Harper and J. B. Rosenbaum (1973). Vegetative stabilization of mill mineral wastes. In: *Ecology and Reclamation of Devastated Land*, Vol. 2 (R. J. Hutnik and G. Davis (eds)), Gordon and Breach, New York, pp. 119–36.

Djuric, D., Z. Kevin, L. Graovac-Leposavic, L. Novak and M. Kop (1971). Environmental contamination by lead from a mine and smelter. *Arch. Environ. Health*. **23**: 275–9.

Douglas, R. J. W. (1970). *Geology and Economic Minerals of Canada. Dept. of Energy, Mines and Resources*, Ottawa. 838 pp.

Down, C. G. (1975). Soil development on colliery waste tips in relation to age. II. Chemical factors. *J. App. Ecol.* **12**: 623–35.

Dubois, L., T. Teichmon, J. M. Airth and J. L. Monkmon (1966). The metal content of urban air. *J. Air Poll. Control Fed.* **16**: 77–8.

Dvorak, A. J. and B. G. Lewis (eds) (1978). Impacts of coal-fired power plants on fish, wildlife, and their habitats. Fish and Wildlife Service, US Dept. of the Interior. Pub. FWS/OBS-78/29. 260 pp.

Ebregt, A. and J. M. A. M. Boldewijn (1977). Influence of heavy metals in spruce forest soil on amylase activity, CO_2 evolution from starch, and soil respiration. *Plant Soil*, **47**: 137–48.

Elseewi, A. A., F. T. Bingham and A. L. Page (1978). Growth and mineral composition of lettuce and Swiss chard grown on fly-ash-amended soils. In: *Environmental Chemistry and Cycling Processes* (D. C. Adriano and I. L. Brisbin (eds)), US Dept. of Energy, Washington, pp. 568–81.

Ernst, W. H. O. (1975). Physiology of heavy metal resistance in plants. In: *Symposium Proceedings, Vol. II. International Conference on Heavy Metals in the Environment* (T. C. Hutchinson (ed)), Toronto, pp. 121–36.

Evans, L. F., N. K. King, D. R. Pockham and E. T. Stephens (1974). Ozone measurements in smoke from forest fires. *Environ. Sci. Technol.* **8**: 75–80.

Fimreite, N. (1970). Mercury uses in Canada and their possible hazards as sources of mercury contamination. *Environ. Poll.* **1**: 119–31.

Fimreite, N., R. W. Fyfe and J. A. Keith (1970). Mercury contamination of Canadian prairie seed eaters and their avian predators. *Can. Field-Nat.* **84**: 269–76.

Forgeron, F. D. (1971). Soil geochemistry in the Canadian Shield. *Can. Mining and Metall. Bull.* 37–42.

Frank, R., H. E. Braun, K. Ishida and P. Suda (1976*a*). Persistent organic and inorganic pesticide residues in orchard soils and vineyards of southern Ontario. *Can. J. Soil Sci.* **56**: 463–84.

Frank, R., K. Ishida and P. Suda (1976*b*). Metals in agricultural soils of Ontario. *Can. J. Soil Sci.* **56**: 181–96.

Freedman, B. (1978). Effects of smelter pollution near Sudbury, Ontario, Canada on surrounding forested ecosystems. Ph.D. Thesis, Department of Botany, University of Toronto, Toronto. 223 pp.

Freedman, B. and T. C. Hutchinson (1979*a*). Contamination of soils and vegetation by heavy metals in Canada. In: *Perspectives in Air Pollution in Canada* (C. Sparrow and L. Foster (eds)), University of Toronto Press (in press).

Freedman, B. and T. C. Hutchinson (1979*b*). Pollutant inputs from the atmosphere and accumulations in soils and vegetation near a nickel–copper smelter at Sudbury, Ontario, Canada. *Can. J. Bot.* **58**: 108–32.

Furr, A. K., A. W. Lawrence, S. S. C. Tong, M. C. Grandolfo, R. A. Hofstader, C. A. Bache, W. H. Gutenmann and D. J. Lisk (1976). Multi-element and chlorinated hydrocarbon analysis of municipal sewage sludges of American cities. *Environ. Sci. Technol.* **10**: 683–7.

Furr, A. K., T. F. Parkinson, R. A. Hinrichs, D. R. van Campen, C. A. Bache, W. H. Gutenmann, L. E. St. John, I. S. Pakkala and D. J. Lisk (1977). National survey of elements and radioactivity in fly ashes. Absorption of elements by cabbage grown in fly ash–soil mixtures. *Environ. Sci. Technol.* **11**: 1194–201.

Gaynor, J. D. and R. L. Halstead (1976). Chemical and plant extractability of metals and plant growth on soils amended with sludge. *Can. J. Soil Sci.* **56**: 1–8.

Gemmell, R. P. (1977). *Colonization of Industrial Wasteland. The Institute of Biology's Studies in Biology No. 80.* Edward Arnold, London. 75 pp.

Godzik, S., T. Florkowski, S. Piorek and M. M. A. Sassen (1979). An attempt to determine the tissue contamination of *Quercus robur* L. and *Pinus sylvestris* L. foliage by particulates from zinc and lead smelters. *Environ. Pollut.* **18**: 97–105.

Goodman, G. T. and T. M. Roberts (1971). Plants and soils as indicators of metals in the air. *Nature.* **231**: 287–92.

Goodman, G. T., C. E. R. Pitcairn and R. P. Gemmell (1973). Ecological factors affecting growth on sites contaminated with heavy metals. In: *Ecology and Reclamation of Devastated Land*, Vol. 2 (R. J. Hutnik and G. Davis (eds)), Gordon and Breach, New York, pp. 149–71.

Granat, L., R. O. Hallberg and H. Rodhe (1975). The global sulfur cycle. In: *Nitrogen, Phosphorus and Sulfur—Global Cycles* (B. H. Svensson and R. Soderlund (eds)), SCOPE Report 7, Ecol. Bull. 22, Stockholm, pp. 89–134.

Greenberg, R. R., W. H. Zoller and G. E. Gordon (1978*a*). Composition and size distributions of particles released in refuse incineration. *Environ. Sci. Technol.* **12**: 566–73.

Greenberg, R. R., G. E. Gordon, W. H. Zoller, R. B. Jacko, D. W. Neuendorf and K. J. Yost (1978*b*). Composition of particulates emitted from the Nicosia Municipal Incinerator. *Environ. Sci. Technol.* **12**: 1829–32.

Gregory, R. P. G. and A. O. Bradshaw (1965). Heavy metal tolerance in populations of *Agrostis tenuis* SIBTH and other grasses. *New Phytol.* **69**: 131–43.

Harris, M. M. and M. F. Jurgensen (1977). Development of *Salix* and *Populus* mycorrhizae in metallic mine tailings. *Plant Soil.* **47**: 509–17.

Hemkes, O. J. and J. Hartmans (1974). Copper content in grass and soil under high-voltage lines in industrial and rural areas. In: *Trace Substances in Environmental Health—VII* (D. D. Hemphill (ed)), University of Missouri, Columbia, pp. 175–8.

Hocking, D., P. Kuchar, J. A. Plumbeck and R. A. Smith (1978). The impact of gold smelter emissions on vegetation and soils of a sub-Arctic forest tundra transition ecosystem. *J. Air Poll. Control Assoc.* **28**: 133–7.

Hogan, G. D., G. M. Courtin and W. E. Rauser (1977*a*). The effects of soil factors on the distribution of *Agrostis gigantea* on a mine waste site. *Can. J. Bot.* **55**: 1038–42.

Hogan, G. D., G. M. Courtin and W. E. Rauser (1977*b*). Copper tolerance in clones of *Agrostis gigantea* from a mine waste site. *Can. J. Bot.* **55**: 1043–50.

Hoggan, M. C., A. Davidson, M. F. Brunelle, J. S. Neuitt and J. D. Gins (1978). Motor vehicle emissions and atmospheric lead concentrations in the Los Angeles area. *J. Air Poll. Control Assoc.* **28**: 1200–6.

Holloway, G. T. (1917). Report of the Ontario Nickel Commission, with Appendix. Legislative Assembly of Ontario. Toronto.

Horton, J. H. and R. S. Dorsett (1976). Savanah River Laboratory, personal communication cited in Turner and Strojan (1978).

Huckabee, J. W. (1973). Mosses: Sensitive indicators of air borne mercury pollution. *Atmos. Environ.* **7**: 749–54.

Humphreys, M. O. and A. D. Bradshaw (1976). Genetic potentials for solving problems of soil mineral stress: Heavy metal toxicities. In: *Proceedings of Workshop on Problems of Soil Mineral Stress*, Beltsville, Maryland, pp. 95–105.

Hunter, J. G. and O. Vergnano (1952). Nickel toxicity in plants. *Ann. Appl. Biol.* **39**: 279–84.

Hutchinson, T. C. (1972). The occurrence of lead, cadmium, nickel, vanadium, and chloride in soils and vegetation of Toronto in relation to traffic density. Institute for Environmental Studies, University of Toronto. Pub. EH-2, Toronto. 27 pp.

Hutchinson, T. C. and L. M. Whitby (1974). Heavy metal pollution in the Sudbury mining and smelting region of Canada. I. Soil and vegetation contamination by nickel, copper, and other metals. *Environ. Conserv.* **1**: 123–32.

Hutchinson, T. C. and L. M. Whitby (1977). The effects of acid rainfall and heavy metal particulates on a boreal forest ecosystem near the Sudbury smelting region of Canada. *Water Air Soil Poll.* **7**: 421–38.

Hutchinson, T. C., M. Czuba and L. Cunningham (1974). Lead, cadmium, zinc, copper, and nickel distributions in vegetables and soils of an intensely cultivated area and levels of copper, lead, and zinc in the growers. In: *Trace Substances in Environmental Health—VIII* (D. D. Hemphill (ed)), University of Missouri, Columbia.

Hutchinson, T. C., W. Gizyn, M. Havas and V. Zobens (1978). Effect of long-term lignite burns on Arctic ecosystems at the Smoking Hills, N.W.T. In: *Trace Substances in Environment Health—XII* (D. D. Hemphill (ed)), University of Missouri, Columbia, pp. 317–32.

Jacko, R. B. and D. W. Neuendorf (1977). Trace metal particulate emission test results from a number of industrial and municipal point sources. *J. Air Poll. Control Assoc.* **27**: 989–94.

Jaffre, T., R. R. Brooks, J. Lee and R. D. Reeves (1976). *Sebertia acuminata*: A hyperaccumulator of nickel from New Caledonia. *Science.* **193**: 579–80.

Jain, S. K. and A. D. Bradshaw (1966). Evolutionary divergence among adjacent plant populations. 1. The evidence and its theoretical analysis. *Heredity.* **21**: 407–41.

Joensuu, O. I. (1971). Fossil fuels as a source of mercury pollution. *Science.* **172**: 1027–8.

John, M. K. (1971). Lead contamination of some agricultural soils in western Canada. *Environ. Sci. Technol.* **5**: 1199–203.

John, M. K., H. H. Chuah and C. J. van Laerhoven (1972). Cadmium contamination of soil and its uptake by oats. *Environ. Sci. Technol.* **6**: 555–65.

John, M. K., C. J. van Laerhoven and C. H. Cross (1975). Cadmium, lead, and zinc accumulation in soils near a smelter complex. *Environ. Lett.* **10**: 25–35.

John, M. K., C. J. van Laerhoven and J. H. Bjerring (1976). Effect of a smelter complex on the regional distribution of cadmium, lead, and zinc in litters and soil horizons. *Arch. Environ. Contam. Toxicol.* **4**: 456–68.

Johnels, A., G. Tyler and T. Westermark (1979). A history of mercury levels in Swedish fauna. *Ambio.* **8**: 160–8.

Johnson, R. D., R. J. Miller, R. E. Williams, C. M. Wai, A. C. Wiese and J. E. Mitchell (1975). The heavy metal problem of Silver Valley, Northern Idaho. In: *Symposium Proceedings, Vol. II—Part 2. International Conference on Heavy Metals in the Environment* (T. C. Hutchinson (ed)), Toronto, Canada, pp. 465–87.

Jordan, M. J. (1975). Effects of zinc smelter emissions and fire on a chestnut–oak woodland. *Ecology.* **56**: 78–91.

Jordan, M. J. and M. P. Lechevalier (1975). Effects of zinc-smelter emissions on forest soil microflora. *Can. J. Microbiol.* **21**: 1855–65.

Keeney, D. R. (1975). Toxic elements in agriculture. Unpublished manuscript. Dept. of Soil Science, University of Wisconsin. Madison, Wisconsin. 32 pp.

Kellogg, W. W., R. D. Cadle, E. R. Allan, A. L. Lazrus and E. A. Martell (1972). The sulfur cycle. *Science.* **175**: 587–95.

Kerin, Z. (1975). Relationship between lead content in the soil and in the plants contaminated by industrial emissions of lead aerosols. In: *Symposium Proceedings, Vol. II—Part 2. International Conference on Heavy Metals in the Environment* (T. C. Hutchinson (ed)), Toronto, Canada, pp. 487–502.

Kilotat, E. J., P. Eng. and H. J. Wilson (1970). An evaluation of air pollution levels in Sydney, Nova Scotia. Air Pollution Control Division, Environmental Health Directorate, Dept. of National Health and Welfare, Ottawa. Pub. APCD 70-1. 14 pp.

Klein, D. H. (1972). Mercury and other metals in urban soil. *Environ. Sci. Technol.* **6**: 560–2.

Klein, D. H. and P. Russell (1973). Heavy metals: Fallout around a power plant. *Environ. Sci. Technol.* **7**: 357–8.

Klein, D. H., A. W. Andren and N. E. Bolton (1975*a*). Trace element discharges from coal combustion for power production. *Water Air Soil Pollut.* **5**: 71–7.

Klein, D. H., A. W. Andren, J. A. Carter, J. F. Emery, C. Feldman, W. Falkerson, W. S. Lyon, J. C. Ogle, Y. Talmi, R. I. van Hook and N. Bolton (1975*b*). Pathways of thirty-seven trace elements through coal-fired power plant. *Environ. Sci. Technol.* **9**: 973–9.

Kruckeberg, A. R. (1954). The ecology of serpentine soils. III. Plant species in relation to serpentine soils. *Ecology*. **35**: 267–74.

Kuja, A. (1979). Reclamation studies of metalliferous tailings from several sites in the Yukon Territory, Canada. M.Sc. Thesis, Dept. of Botany, University of Toronto, Toronto. 200 pp.

Lagerwerff, J. V. and A. W. Specht (1970). Contamination of roadside soil and vegetation with cadmium, nickel, lead, and zinc. *Environ. Sci. Technol.* **4**: 583–6.

Lagerwerff, J. V., G. T. Biersdorf and D. L. Brower (1976). Retention of metals in sewage sludge. I. Constituent heavy metals. *J. Environ. Qual.* **5**: 19–22.

Landrigan, P. J. *et al.* (1975). Epidemic lead absorption near an ore smelter. *New England J. Med.* **292**: 123–9.

Law, S. L. and G. E. Gordon (1979). Sources of metals in municipal incinerator emissions. *Environ. Sci. Technol.* **13**: 432–8.

Leland, H. V., E. D. Copenhaver and L. S. Corrill (1974). Heavy metals and other trace elements. *Journal W.P.C.F.* **46**: 1452–76.

Leroy, J. C. and H. Keller (1972). How to reclaim mined areas, tailings ponds, and dumps into valuable land. *World Mining*. 34–41.

Lewis, W. M. (1975). Effects of forest fires on atmospheric loads of soluble nutrients. In: *Mineral Cycling in Southeastern Ecosystems* (F. G. Howell, J. B. Gentry and M. H. Smith (eds)), Pub. Technical Information Center, Office of Public Affairs, US Energy Research and Development Administration, pp. 833–46.

Linzon, S. N., B. L. Chai, P. J. Temple, R. G. Pearson and M. L. Smith (1976). Lead contamination of urban soils and vegetation by emissions from secondary lead smelters. *J. Air Poll. Control Assoc.* **26**: 650–4.

Little, P. and H. M. Martin (1972). A survey of zinc, lead, and cadmium in soil and natural vegetation around a smelting complex. *Environ. Pollut.* **3**: 241–54.

Little, P. and M. H. Martin (1974). Biological monitoring of heavy metal pollution. *Environ. Pollut.* **6**: 1–20.

McGovern, P. C. (1976). Air quality assessment studies in the Wawa area, 1975. Ontario Ministry of the Environment, Northeastern Region, Sudbury. 51 pp.

McGovern, P. C. and D. Balsillie (1972). Sulfur dioxide levels and environmental studies in the Wawa area during 1971. Air Quality Branch, Ontario Ministry of the Environment. Sudbury. 38 pp.

MacLean, A. J. and A. J. Dekker (1978). Availability of zinc, copper and nickel to plants grown in sewage-treated soils. *Can J. Soil Sci.* **58**: 381–9.

MacLean, A. J., B. Store and W. B. Cordukes (1973). Amounts of mercury in soil of some golf course sites. *Can. J. Soil Sci.* **53**: 130–2.

Malyuga, D. P. (1964). *Biogeochemical Methods of Prospecting*. Academic Science Press, Moscow. Translated Consultants Bureau, New York. 205 pp.

Manson, A. N. and H. W. De Koning (1973). A look at lead emissions in Vancouver, Toronto, Montreal. *Water Poll. Control.* 70–3.

Martens, D. C. and B. R. Beahm (1978). Chemical effects on plant growth of fly-ash incorporation into soil. In: *Environmental Chemistry and Cycling Processes* (D. C. Adriano and I. L. Brisbin (eds)), Technical Information Centre, US Dept. of Energy, Washington, pp. 637–44.

Massey, H. F. (1972). pH and soluble Cu, Ni, and Zn in eastern Kentucky coal mine spoil materials. *Soil Sci.* **114**: 217–21.

Massey, H. F. and R. I. Barnhisel (1972). Copper, nickel and zinc released from acid coal mine spoil materials of eastern Kentucky. *Soil Sci.* **113**: 207–12.

Michelutti, R. E. (1974). How to establish vegetation on high iron–sulfur mine tailings. Falconbridge Nickel Mines Ltd, Sudbury, Ontario. 6 pp.

Miles, J. R. W. (1968). Arsenic residues in agricultural soils of Southwestern Ontario. *J. Agric. Food Chem.* **16**: 620–2.

Mishra, D. and M. Kar (1974). Nickel in plant growth and metabolism. *Bot. Rev.* **40**: 395–452.

Moore, T. R. and R. C. Zimmerman (1977). Establishment of vegetation on serpentine asbestos mine wastes, southeastern Quebec, Canada. *J. Appl. Ecol.* **14**: 589–99.

Motto, H. L., R. H. Daines, D. M. Chilko and C. K. Motto (1970). Lead in soils and plants: Its relationship to traffic volume and proximity to highways. *Environ. Sci. Technol.* **4**: 231–7.

Munn, R. E., D. A. Thomas and A. F. W. Cole (1969). A study of suspended particulate and iron concentrations in Windsor, Ontario. *Atmos. Environ.* **3**: 1–10.

Natusch, D. F. S., C. F. Bauer, H. Matusiewicz, C. A. Evans, J. Baker, A. Loh, R. W. Linton and P. K. Hopke (1975). Characterization of trace elements in fly ash. In: *Symposium Proceedings, Vol. II—Part 2. International Conference on Heavy Metals in the Environment* (T. C. Hutchinson (ed)), Toronto, Canada, pp. 553–78.

Nieboer, E., H. M. Ahmed, K. J. Puckett and D. H. S. Richardson (1972). Heavy metal content of lichens in relation to distance from a nickel smelter in Sudbury, Ontario. *Lichenol.* **5**: 292–304.

Oldreive, C. P. (1976). Sydney area ambient air monitoring annual report 1976. Inspection and Monitoring Division, Nova Scotia Department of the Environment, Sydney. 30 pp.

Oliver, B. G. and J. Kinrade (1972). Heavy metal concentrations in Ottawa River and Rideau River sediments. Inland Waters Branch, Dept. of the Environment. Scientific Series, Pub. No. 14, Ottawa. 10 pp.

Ondov, J. M., R. C. Ragaini and A. H. Biermann (1979*a*). Elemental emissions from a coal-fired power plant. Comparison of a Venturi wet scrubber system with a cold-sided electrostatic precipitator. *Environ. Sci. Technol.* **13**: 598–607.

Ondov, J. M., R. C. Ragaini and A. H. Biermann (1979*b*). Emissions and particle-size distributions of minor and trace elements at two western coal-fired power plants equipped with cold-side electrostatic precipitators. *Environ. Sci. Technol.* **13**: 947–53.

Paciga, J. J. (1975). Trace element characterization and size distributions of atmospheric particulate matter in Toronto. Ph.D. Thesis, Dept. of Chem. Eng. and Applied Chem., University of Toronto, Toronto. 200 pp.

Page, A. L. (1974). Fate and effects of trace elements in sewage sludge when applied to agricultural lands, Pub. EPA-670/2-74-005. Office of Research and Development, US Environmental Protection Service, Cincinnati, Ohio. 97 pp.

Page, A. L. and F. T. Bingham (1973). Cadmium residues in the environment. *Residue Rev.* **48**: 1–44.

Page, A. L., T. J. Ganje and M. S. Joshi (1971). Lead quantities in plants, soil, and air near some major highways in Southern California. *Hilgardia.* **41**: 1–31.

Persson, H. (1948). On the discovery of *Merceya ligulata* in the Azores with a discussion of the so-called copper mosses. *Rev. Bryol. Lichen.* **17**: 75–8.

Peterson, P. J. (1975). Element accumulation by plants and their tolerance of toxic mineral soils. In: *Symposium Proceedings, Vol. II, International Conference on Heavy Metals in the Environment* (T. C. Hutchinson (ed)), Toronto, Canada, pp. 39–59.

Peterson, H. B. and R. F. Nielson (1973). Toxicities and deficiencies in mine tailings. In: *Ecology and Reclamation of Devastated Land*, Vol. 1 (R. J. Hutnik and G. Davis (ed)), Gordon and Breach, New York, pp. 15–24.

Pilegaard, K. (1976). Heavy metals in bulk precipitation, soil, and vegetation around a power plant. In: *Proceedings of the Kuopio Meeting on Plant Damages Caused by Air Pollution* (L. Karenlampi (ed)), Kuopio, pp. 46–56.

Proctor, J. (1971*a*). The plant ecology of serpentine. II. Plant response to serpentine soils. *J. Ecol.* **59**: 397–410.

Proctor, J. (1971*b*). The plant ecology of serpentine. III. The influence of a high magnesium/calcium ratio and high nickel and chromium levels in some British and Swedish serpentine soils. *J. Ecol.* **59**: 827–42.

Proctor, J. and S. R. J. Woodell (1975). The ecology of serpentine soils. *Adv. Ecol. Res.* **9**: 255–365.

Purves, D. (1972). Consequences of trace-element contamination of soils. *Environ. Pollut.* **3**: 17–24.

Radke, L. F., J. L. Stith, D. A. Hegg and P. V. Hobbs (1977). Airborne studies of particles and gases from forest trees. *J. Air Poll. Control Assoc.* **28**: 30–4.

Ragaini, R. C., H. R. Ralston and N. Roberts (1977). Environmental trace metal contamination in Kellogg, Idaho, near a lead smelting complex. *Environ. Sci. Technol.* **11**: 773–81.

Reay, R. F. (1972). The accumulation of arsenic from arsenic-rich natural waters by aquatic plants. *J. Appl. Ecol.* **9**: 557–65.

Reilly, C. (1967). Accumulation of copper by some Zambian plants. *Nature.* **215**: 666–9.

Reilly, A. and C. Reilly (1973). Copper-induced chlorosis in *Becium homblei*. *Plant Soil.* **38**: 671–4.

Roberts, T. M. (1975). A review of some biological effects of lead emissions from primary and secondary smelters. In: *Symposium Proceedings, Vol. II—Part 2. International Conference on Heavy Metals in the Environment* (T.C. Hutchinson (ed)), Toronto, Canada, pp. 503–32.

Roberts, T. M. and G. T. Goodman (1974). The persistence of heavy metals in soils and natural vegetation following closure of a smelter. In: *Trace Substances in Environmental Health—VIII* (D. D. Hemphill (ed)), University of Missouri, Columbia, pp. 117–25.

Roberts, T. M., T. C. Hutchinson, J. Paciga, A. Chattopadhyay, R. E. Jervis, J. van Loon and D. K. Parkinson (1974*a*). Lead contamination around secondary smelters: Estimation of dispersion and accumulation by humans. *Science*. **186**: 1120–3.

Roberts, T. M., J. Paciga, T. C. Hutchinson, R. E. Jervis, A. Chattopadhyay, J. C. van Loon and F. Huhn (1974*b*). Lead contamination around two secondary smelters in downtown Toronto—Estimation of ongoing pollution and accumulation by humans. Institute for Environmental Studies, University of Toronto, Toronto. Pub. EE-1. 74 pp.

Roberts, T. M., W. Gizyn and T. C. Hutchinson (1975). Lead contamination of air, soil, vegetation and people in the vicinity of secondary lead smelters. In: *Trace Substances in Environmental Health—VIII* (D. D. Hemphill (ed)), University of Missouri, Columbia, pp. 155–66.

Rorison, I. H. (1973). The effect of extreme soil acidity on the nutrient uptake and physiology of plants. In: *Acid Sulfate Soil. Proc. Int. Symp. on Acid Sulfate Soils*, Pub. 18, Vol. 1 (H. Dost (ed)), Int. Inst. Land Reclamation and Improvement, Wageningen, pp. 223–51.

Ruhling, A. and G. Tyler (1968). An ecological approach to the lead problem. *Bot. Notiser*. **121**: 321–42.

Ruhling, A. and G. Tyler (1973). Heavy metal pollution and decomposition of spruce needle litter. *Oikos*. **24**: 402–16.

Rutherford, G. K. and C. R. Bray (1979). Extent and distribution of soil heavy metal contamination near a nickel smelter at Coniston, Ontario. *J. Environ. Qual*. **8**: 219–22.

Schmitt, N., E. L. Devlin, A. A. Larsen, E. D. McCausland and J. M. Saville (1971). Lead poisoning in horses. *Arch. Environ. Health*. **23**: 185–95.

Schwitzgebel, K., F. B. Meserole, R. G. Oldham, R. A. Magee, F. G. Mesich and T. L. Thoem (1975). Trace element discharge from coal-fired power plants. In: *Symposium Proceedings, Vol. II—Part 2. International Conference on Heavy Metals in the Environment* (T. C. Hutchinson (ed)), Toronto, Canada, pp. 533–52.

Scoggan, H. J. (1950). The flora of Bic and the Gaspe Peninsula, Quebec. National Museum of Canada, Bulletin No. 115, Biological Series No. 39, Ottawa. 399 pp.

Seto, P. and P. Deangelis (1978). Concepts of sludge utilization on agricultural land. In: *Sludge Utilization and Disposal Conference. Proceedings No. 6*, Environment Canada, Ottawa, pp. 138–155.

Shacklette, H. T. (1965). Bryophytes associated with mineral deposits and solutions in Alaska. Geol. Survey Bull. 1198C, Washington, D.C.

Sidle, R. C., J. E. Hook and L. T. Kardos (1976). Heavy metals application and plant uptake in a land disposal system for waste water. *J. Environ. Qual*. **5**: 97–102.

Siegel, B. Z. and S. M. Siegel (1978). Mercury emission in Hawaii: Aerometric study of the Kalalua Eruption of 1977. *Environ. Sci. Technol*. **12**: 1036–9.

Smith, M. J. (1974). Acid production in mine drainage streams. In: *Extraction of Minerals and Energy: Today's Dilemmas*. (R. A. Deju (ed)), Ann Arbor Science Pub. Inc., Ann Arbor, Michigan, pp. 57–75.

Smith, R. A. H. and A. D. Bradshaw (1979). The use of metal-tolerant plant populations for the reclamation of metalliferous waste. *J. Appl. Ecol.* **16**: 595–612.

Smith, R. D., J. A. Campbell and K. K. Nielson (1979). Concentration dependence upon particle size of volatilized elements in fly ash. *Environ. Sci. Technol.* **13**: 553–8.

Soane, B. O. and D. H. Saunder (1959). Nickel and chromium toxicity of serpentine soils in southern Rhodesia. *Soil Sci.* **88**: 322–30.

Stoiber, R. E. (1973). Sulfur dioxide contributions to the atmosphere by volcanoes. *Science*, **182**: 577–8.

Stone, E. L. and V. R. Timmer (1975). On the copper content of some northern conifers. *Can. J. Bot.* **53**: 1453–6.

Strojan, C. L. (1978*a*). The impact of zinc smelter emissions on forest litter arthropods. *Oikos*. **31**: 41–6.

Strojan, C. L. (1978*b*). Forest leaf litter decomposition in the vicinity of a zinc smelter. *Oecologia*. **32**: 203–12.

Struthers, P. H. (1964). Chemical weathering of strip-mine soils. *Ohio J. Sci.* **64**: 125–131.

Tejning, S. (1967). Mercury in pheasant (*Phasianus colchicus* L.) deriving from seed grain dressed with methyl and ethyl mercury compounds. *Oikos*. **18**: 334–44.

Temple, P. J. (1975). Phytotoxicity surveys in the vicinity of International Nickel Co., Port Colbourne—1975. Phytotoxicology Section, Air Management Branch, Ontario Ministry of the Environment, Toronto. 19 pp.

Temple, P. J. (1978*a*). Phytotoxicology surveys in the vicinity of International Nickel Co., Port Colbourne—1977. Phytotoxicology Section, Air Management Branch, Ontario Ministry of the Environment, Toronto. 13 pp.

Temple, P. J. (1978*b*). Investigations of the effect of heavy metals on muck farms east of International Nickel Co., Port Colbourne—1976–77. Phytotoxicology Section, Air Management Branch, Ontario Ministry of the Environment, Toronto. 26 pp.

Thornton, I. (1975). Some aspects of environmental geochemistry in Britain. In: *Symposium Proceedings, Vol. II—Part 1. International Conference on Heavy Metals in the Environment* (T.C. Hutchinson (ed)), Toronto, Canada, pp. 17–38.

Tomassini, F. D., K. J. Puckett, E. Nieboer, D. H. S. Richardson and B. Grace (1976). Determination of copper, iron, nickel, and sulfur by X-ray fluorescence in lichens from the Mackenzie Valley, Northwest Territories, and the Sudbury District, Ontario. *Can. J. Bot.* **54**: 1591–1603.

Townsend, W. N. and E. W. F. Gillham (1973). Pulverised fuel ash as a medium for plant growth. In: *The Ecology of Resource Degradation and Renewal* (M. J. Chadwick and G.T. Goodman (eds)), Blackwell Sci. Pub., London, pp. 287–304.

Turner, F. B. and C. L. Strojan (1978). Coal combustion, trace element emissions, and mineral cycles. In: *Environmental Chemistry and Cycling Processes* (D. C. Adriano and I. L. Brisbin, Jr (eds)), Technical Information Center, US Dept. of Energy, Washington, pp. 34–58.

Tyler, G. (1974). Heavy metal pollution and soil enzymatic activity. *Plant Soil*. **41**: 303–11.

Tyler, G. (1975). Heavy metal pollution and mineralization of oxygen in forest soils. *Nature*. **255**: 701–2.

Tyler, G. (1976). Heavy metal pollution, phosphatase activity, and mineralization of organic phosphorus in forest soils. *Soil Biol. Biochem.* **8**: 327–32.

Tyler, G. (1979). Heavy metal pollution and soil enzymatic activity. *Plant Soil*. **41**: 303–11.

van Horn, W. H. (1975). Materials balance and technology assessment of mercury and its compounds on a national and regional basis. Prepared by U.R.S. Reşearch Co. for US environmental Protection Agency, EPA 560/3-75-007.

van Loon, J. C. (1973). Toronto's precipitation analyzed for heavy metal content. *Water Poll. Cont.* 38–41.

van Loon, J. C. (1974). Analysis of heavy metals in sewage sludge and liquids associated with sludges. Presented at Canada/Ontario Sludge Handling and Disposal Seminar, Toronto, Sept. 18–19, 1974. 17 pp.

van Loon, J. C. and R. J. Beamish (1977). Heavy metal contamination by atmospheric fallout of several Flin Flon area lakes and the relation to fish populations. *J. Fish Res. Board Can.* **34**: 899–906.

Vandegrift, A. E., L. J. Shannon, E. E. Sallee, P. G. Garman and W. R. Park (1971). Particulate air pollution in the United States. *J. Air Poll. Control Assoc.* **21**: 321–8.

Vaughan, B. E. *et al.* (1975). Review of potential impact on health and environmental quality from metals entering the environment as a result of coal utilization. Report NP-20585, Pacific Northwest Laboratories, Battelle Memorial Institute, Richland, Washington. Cited in Turner and Strojan (1978).

Vergnano, O. and J. G. Hunter (1952). Nickel and cobalt toxicities in oat plants. *Ann. Bot.* **17**: 317–28.

Ward, N. I., R. R. Brooks, E. Roberts and C. R. Boswell (1977). Heavy metal pollution from automotive emissions and its effect on roadside soils and pasture species in New Zealand. *Environ. Sci. Technol.* **11**: 917–23.

Warren, H. V., R. E. Delavault and J. Barakso (1966*a*). Some observations on the geochemistry of mercury as applied to prospecting. *Econ. Geol. Ser. Can.* **61**: 1010–28.

Warren, H. V., R. E. Delavault and C. H. Cross (1966*b*). Mineral contamination in soil and vegetation and its possible relation to public health. Canadian Council of Resource Ministers—Background Papers for National Conference on Pollution and Our Environment, Montreal. 11 pp.

Warren, H. V., R. E. Delavault and C. H. Cross (1969). Base metal pollution in soils. In: *Trace Substances in Environmental Health—III* (D. D. Hemphill (ed)), University of Missouri, Columbia, pp. 9–19.

Warren, H. V., R. E. Delavault and K. W. Fletcher (1971*a*). Metal pollution—a growing problem in industrial and urban areas. *Can. Mining and Metall. Bull.* 34–45.

Warren, H. V., R. E. Delavault, K. Fletcher and E. Wilks (1971*b*). A study in lead pollution. *Western Miner*. 22–6.

Watson, A. P. (1976). Trace element impact on forest floor litter in the New Lead Belt Region of southeastern Missouri. In: *Trace Substances in Environmental Health—IX* (D. D. Hemphill (ed)), University of Missouri, Columbia, pp. 227–36.

Watson, A. P., R. I. van Hook, D. R. Jackson and D. E. Reichle (1976). Impact of a lead mining–smelting complex on the forest floor litter arthropod fauna in the New Lead Belt Region of southeast Missouri. Oak Ridge National Laboratory, Environmental Sciences Division ·Pub. Nò 881, Oak Ridge, Tennessee.

Watson, W. Y. and D. H. Richardson (1972). Appreciating the potential of a devastated land. *For. Chron.* 312–15.

Webb, J. S., P. L. Lowenstein, R. J. Howarth, I. Nichol and R. Foster. (1973). Provisional geochemical atlas of Northern Ireland. Applied Geochemistry Research Group, Technical Communication No. 60. Imperial College of Science and Technology, London.

Webber, L. R. and E. G. Beauchamp (1975). Heavy metals in corn grown on waste-amended soils. In: *Symposium Proceedings, Vol. II—Part 1. International Conference on Heavy Metals in the Environment* (T. C. Hutchinson (ed)), Toronto, Canada, pp. 443–52.

Wheeler, G. L. and G. L. Rolfe (1979). The relationship between daily traffic volume and the distribution of lead in roadside soil and vegetation. *Environ. Pollut.* **18**: 265–74.

Whitby, L. M. and T. C. Hutchinson (1974). Heavy metal pollution in the Sudbury mining and smelting region of Canada. II. Soil toxicity tests. *Environ. Conserv.* **1**: 191–200.

Whitby, L. M., A. J. MacLean, M. Schnitzer and J. D. Gaynor (1977). Sources, storage, and transport of heavy metals in agricultural watersheds. International Reference Group on Great Lakes Pollution from Land Use Activities, International Joint Commission, Windsor, Ontario. 142 pp.

Whittaker, R. H. (1954). The ecology of serpentine soils. IV. The vegetational response to serpentine soils. *Ecology.* **35**: 275–88.

Winterhalder, K. (1978). A historical perspective of mining and reclamation in Sudbury. Proceedings of the Third Annual Meeting, Canadian Land Reclamation Association, Sudbury, May 29–June 1, 1978. 13 pp.

Wixson, B. G. and J. H. Arvik (eds) (1973). An interdisciplinary investigation of environmental pollution by lead and other heavy metals from industrial development in the New Lead Belt of southeastern Missouri. University of Missouri, Rolla and Columbia.

Wolfe, W. J. (1971). Biogeochemical prospecting in glaciated terrain of the Canadian Precambrian Shield. *Can. Mining and Metall. Bull.* 72–80.

Wood, C. W. and T. N. Nash III (1976). Copper smelter effluent effects on Sonoran Desert vegetation. *Ecology.* **57**: 1311–16.

Zenz, D. R., J. Peterson and C. Lue-Hing (1978). United States sludge disposal regulations—impacts upon municipal sewage treatment agencies—comparison with experimental results. In: *Sludge Utilization and Disposal Conference. Proceedings No. 6*, Environment Canada, Ottawa, pp. 297–334.

CHAPTER 3

Cycling of Trace Metals in Ecosystems

M: K. Hughes
Department of Biology, Liverpool Polytechnic, UK

1. INTRODUCTION

Terrestrial ecosystems receive inorganic materials from the atmosphere and from their geological substrates. In undisturbed ecosystems the principal source of major nutrient cations is weathering of the materials underlying the ecosystem itself and of the rocks and soils to which it is connected by the hydrological cycle. In many cases ecosystems lose more calcium, magnesium, sodium and potassium to stream water output than they gain from the atmosphere by rainfall or other inputs (Likens *et al.*, 1977). Even though such transfers may be small compared to the ecosystem's total nutrient stock, it is reasonable to assume that it may only be sustained if a quantity of nutrient cations is released from the substrate by weathering to replace some of the materials lost. The limited data available support this assumption (Bormann and Likens, 1979; Woodwell *et al.*, 1975). Just as for major nutrient cations, the availability of trace metals to undisturbed ecosystems is determined largely by the balance of weathering release and hydrological loss.

Although the geological substrate is the largest source of major nutrient cations in undisturbed ecosystems, the atmosphere may be a significant, if smaller source (Likens *et al.*, 1977). The input of both major nutrients and trace metals from the atmosphere will reflect, *inter alia*, the burden of dissolved and particulate material in the ambient air. This will vary spatially according to the proximity of, for example, the sea or deserts. It will vary temporally as phenomena such as natural fire or volcanic activity occur, bringing with them changes in the natural metal content of the atmosphere.

On a longer time scale, processes such as degassification of the earth's crust will have an effect.

In the last few millenia and more particularly in the last two centuries, a major new factor has emerged in the transfer of trace metals in the environment. Emissions of certain elements to the atmosphere as a result of human activity now rival or exceed the scale of natural emissions. Nriagu (1979) calculates natural and man-effected emissions in millions of kg/yr to be 0·83 and 7·3 for cadmium, 24·5 and 449 for lead, 43·5 and 314 for zinc, 18·5 and 56 for copper and 26 and 47 for nickel, respectively. Many different activities are involved in producing such emissions, the particular physical and chemical nature of the emission varying with the source. Hutchinson and Freedman (Chapter 2) give an account of these many sources and their associated emissions. The majority of emissions of interest are as particles which are subsequently deposited dry or in precipitation onto an ecosystem. The greatest rates of deposition usually occur nearest the pollution source. Air concentration and rates of deposition tend to show an exponential decrease with increasing distance from the pollution source. The physical and chemical nature of the particles and their size interact with atmospheric conditions to determine their atmospheric residence time. This in turn determines the distance the particles travel from their source. In general, larger, heavier particles are deposited first and so only travel short distances, whilst lighter, smaller particles travel further. Consequently, changes in atmospheric input of trace metals to ecosystems tend to be most apparent close to pollution sources producing relatively large particles, for example, smelting and foundry operations (Roberts, 1972; Boggess and Wixson, 1977). These and other activities also produce particles containing trace metals in one or another chemical form which are so small that they have atmospheric residence times of as much as one month (Bowen, 1975). This is particularly the case for lead-containing particles emitted by motor vehicles using fuel containing alkyl lead anti-knock compounds. Chamberlain *et al.* (1978) report diameters (MMD) as low as 0·03 μm for lead-containing particles emitted from cars at cruising speed on a motorway.

Consequently, trace metal contaminants are to be found not only close to major pollution sources, such as foundries or roads, but also in areas distant from these sources. For example, Semb (1978) found elevated levels of lead, cadmium, antimony, arsenic, selenium, vanadium and mercury deposition from the atmosphere over a large part of Norway. Similarly, Rühling and Tyler (1969) presented botanical evidence of a widespread dispersal of heavy metals in Sweden beyond industrial or urban areas.

Siccama and Smith (1978) found an elevated rate of deposition of lead on a watershed in rural New Hampshire, whilst Elias *et al.* (1975) calculated on geochemical grounds that such elevated levels of deposition occurred in a remote mountain region in California. Analysis of plant materials alive at various times in the historic epoch in comparison with material collected recently from the same regions has provided information on long-term trends in deposition. Rühling and Tyler (1968, 1969) used herbarium specimens of Swedish mosses in this way and concluded that there is a greater deposition of lead, copper, zinc and nickel in regions remote from roads or industry than was the case two hundred years ago. Livett *et al.* (1979) report analyses of trace metal contents of peat profiles in Great Britain which, in the case of a site remote from local pollution as well as those in mining areas, indicate an increase in deposition of lead during the past two centuries. It is of interest that their peat profiles from mining areas showed declining lead contents in peat formed after local mining and smelting activity ceased.

Thus, the atmospheric input of trace metals to ecosystems has increased in the last two centuries, not only in areas close to pollution sources but also in more remote areas. The uptake, internal transfer, retention and loss of some of these elements by ecosystems will be examined in this chapter.

2. APPROACHES

2.1. Background

A number of different, but essentially complementary, approaches have been made to the study of the cycling of elements in ecosystems. In what may be termed the 'reductionist' approach, individual transfers are identified and studied either experimentally or observationally. This is often directed toward the formulation and testing of physiological explanations for ecological phenomena. In isolation this could lead to great research effort being expended on transfers of but minor significance in the field. An assessment of the significance of various transfers may be made on the basis of the second or 'ecosystem audit' approach, in which the distribution of an element within the ecosystem is quantified and estimates made of the scale of transfers between major components of the ecosystem. This approach suffers from the disadvantage that most ecosystems studied do not have clear limits. This means that an external and independent check on input and output to the system is difficult to make. In addition to its implications for the reliability of the 'audit', this limitation means that the role of inter-ecosystem transfers, particularly in the hydrological cycle, may not be gauged accurately. In order to overcome these difficulties, a number of studies have been made of the relatively well-defined ecosystem of a self-contained watershed in which atmospheric input and stream water outflow represent, along with *in situ* weathering release, the only significant input and output of elements. Atmospheric input and stream water output are, in principle, eminently measurable quantities. The independent measurement of weathering release presents more problems, but these may be largely overcome in the case of some watersheds. In such cases, the analysis of the balance of watershed input and output may be combined with the audit of ecosystem stocks and internal transfers of an element of interest. This puts the observational basis of our understanding of elemental cycles in and between ecosystems on a much sounder, quantitative basis. In addition, the watershed approach lends itself to field experimentation, in that the watershed may be seen as a finite unit of landscape which may be replicated. Thus the consequences of alternative land mangement practices for intra- and inter-ecosystem element cycling may be investigated by field experimentation. The opportunities for such large-scale field experimentation are, however, limited. In response to this a variant of the watershed approach, the experimental ecological microcosm, has been developed. These are essentially miniature calibrated watersheds in tubs or lysimeters, usually one square metre or less in area, so arranged that all inputs and outputs may be measured. Given their small size and replicability these are particularly useful for experimental purposes.

The descriptive ecosystem audit and watershed approaches, along with experimental variants of the latter, both full scale and in microcosm, generate questions many of which may only be answered by reference to the first 'reductionist' approach. All of this applies equally to the major nutrient cations and to trace metals. In the case of trace metal cycling all these major approaches have been used, namely, the 'reductionist', ecosystem audit and watershed approaches. The application of the microcosm variant of the watershed approach has begun in recent years.

2.2. Needs

The changes in the scale and sources of atmospheric input of trace metals to terrestrial ecosystems discussed in the Introduction to this article have acted as a stimulus to the scientific investigation of the cycling of trace metals in ecosystems. In some cases, work on this topic has been done in a search for explanations of perceived environmental damage. In others attempts are increasingly being made to gain enough understanding of the cycling of particular trace metals to make reliable predictions of the future occurrence of toxic levels in ecosystem components of interest. As will be discussed below, a number of trace metals may occur in forms (chemical species) which may be particularly persistent in organisms or the soil. It is thus possible that trace metals, present in such small quantities in many ecosystems as to be functionally insignificant, may accumulate over a period of years or decades until they reach toxic levels in some organisms.

It has been pointed out (Hall *et al.*, 1975; Hughes *et al.*, 1980) that Smith's (1974) formulation of the relationship between air pollution and temperate forest ecosystems is particularly apposite to the question of trace metal cycling. He defined three classes of impact of air pollution on terrestrial ecosystems. In Class I the biota act as an unaffected sink for pollutants. In Class II intermediate dosages produce sublethal adverse effects on individuals or populations. As a consequence process rates may be changed, particularly in soil. In Class III, high dosages produce lethal effects on individuals or populations, probably disrupting ecosystem processes and reducing structure. A number of Class III occurrences involving trace metals have been reported in the proximity of heavy metal smelters (e.g. Jordan, 1975), although impact may be produced by a complex interaction of trace metals and other pollutants such as sulphur dioxide (Whitby and Hutchinson, 1974). Smith (1974) emphasized our lack of knowledge of Classes I and II. Given the widespread enhancement of the atmospheric input of trace metals to ecosystems referred to in the Introduction, it is important to know what atmospheric input for what time is required to increase trace metal levels in some ecosystem component to a point where the ecosystem passes from Class I to Class II. This requires not only a knowledge of the susceptibility of key ecosystem processes to trace metal toxicity (see Martin and Coughtrey, Chapter 4), but also of the uptake, partition, retention and loss of particular trace metals by ecosystems. If a situation arises where a trace metal accumulates to a greater extent in one ecosystem component than in others, toxic levels may be reached earlier in that component, potentially transferring the ecosystem from Class I to II or from Class II to III. Thus an understanding of the cycling of trace metals in ecosystems is fundamental to an understanding and prediction of their input, particularly outside areas of particularly intense contamination ('hot spots'). A consideration of Smith's (1974) classification leads to a particular emphasis on differential transfer of trace metals between and within ecosystem components, particularly where this might lead to differential accumulation.

In subsequent sections of this chapter the findings of the three main approaches described in Section 2.1 will be described in turn, bearing in mind the needs outlined above.

3. PROCESSES IN TRACE METAL CYCLING

3.1. Entry of Trace Metals into Ecosystems

Weathering release and subsequent soil transfer of trace metals constitute the major source of such elements in undisturbed ecosystems. These processes are reviewed by Thornton (Chapter 1) who also examines their relationship with man-effected inputs. Of these, the most notable change of importance for the landscape at large has been the increasing atmospheric burden and deposition of trace metals associated with urban and industrial activity in recent centuries. It should be remembered that other changes have also taken place in the atmosphere, particularly in the concentration and dispersion of the oxides of sulphur and nitrogen, and of ammonia. These, interacting with micrometeorological conditions may modify the form and composition of trace-metal-containing particles. These factors may in turn affect the relative proportions of trace metal deposition in the dry and wet forms.

Hughes *et al.* (1980) point out that four major factors regulate the access of particles to surfaces in terrestrial ecosystems. These are: particle size, morphology of the deposition surface, aerosol age and wind speed. Particle size is not the simple variable it might seem, as bulk densities may vary greatly, some particles being hollow. Many of the trace-metal-containing particles in the atmosphere are smaller than 10 μm diameter; many smaller than 1 μm (Daines *et al.*, 1970; Harrison *et al.*, 1978; Lundgren, 1971; Robinson and Ludwig, 1964; Lee and von Lehmden, 1973; Melton *et al.*, 1973; Moran *et al.*, 1972; Chamberlain, 1967; Chamberlain *et al.*, 1978). In the smallest particles (diameter $< 1\ \mu$m) Chamberlain (1967) has shown eddy diffusion processes to be the principal route of deposition on surfaces. He showed that, for particles from 1 to 5 μm in diameter leaf surface morphology was most important. Little and Wiffen (1977) report that leaf morphology and particle size may govern deposition rates. Wedding *et al.* (1975) showed much greater deposition of particles on pubescent *Helianthus* leaves than on glabrous *Liriodendron* leaves. Little and Wiffen (1977) explained the greater deposition velocities of fresh aerosols compared to old aerosols in terms of the aggregation of the aged particles.

The configuration of the vegetation as a whole, in its topographic context, will modify wind speed (Grace, 1977) and then the scale of deposition will be in part determined by the four factors described above. Brown (1974) reports broadly similar inputs of potassium and magnesium to a woodland from aerosols and rainfall, whilst the rainfall contribution of calcium was roughly four times as great as that from aerosols. Randall *et al.* (1979) found aerosol and rainfall inputs in rural situations to be similar for iron, but rainfall was more important for zinc, lead, chromium, copper, cadmium and manganese.

Little attention has been paid to the inhalation of airborne trace metals by animals other than man (Quarles *et al.*, 1974). Given that some trace metals occur in sub-micron airborne particles and that even lead may be in relatively soluble forms in these particles (Chamberlain *et al.*, 1978; Biggins and Harrison, 1978) it is possible to argue that, at least among homoiotherms, the smaller the animal the greater the relative importance of inhalation. As Hughes *et al.* (1980) point out, the ratio between respiratory surface area through which dissolved trace metals may pass and body volume will increase with decreasing body size. In addition, lung ventilation rates may be higher than in large mammals leading to a greater effective exposure to atmospheric contamination in small homoiotherms. The greater relative food consumption in these animals will also increase their exposure to trace contaminants. However, since data on these effects are lacking, they cannot be considered in a general account of trace metal cycling in ecosystems.

3.2. Uptake of Trace Metals by Plants

Plants may receive trace metals from either their above-ground surfaces, their roots or some combination of the two.

3.2.1. Above-ground surfaces

Once a trace-metal-containing particle has landed on a leaf or stem surface, the trace metal must be brought into solution in some form before it can gain access to the free space of the peripheral aerial tissues, in the case of vascular plants, or be available for adsorption to bryophyte surfaces or to bark. Whether such solubilization occurs, and its rate, will be determined by the nature and age of the particle and the conditions of micro-environment and chemistry to be found in the intermittent water film of the plant surface. Transfer of trace metals into the plant from solution on and near its surface is discussed in Volume 1. Absorbed metal has several alternative fates in the leaf or bark free space. The trace metal may be bound in the apoplast and lost along with undissolved topical particles when the plant part is shed. Loss by leaching may occur (Tukey *et al.*, 1958; Tukey,

1971; Horler and Barber, 1979). The trace metal may penetrate the leaf symplast, or it may cross a membrane, enter a sieve element and be moved away from the site of entry through the phloem. That this can occur even for lead has been demonstrated experimentally (Haghiri, 1974; Hemphill and Rule, 1975; Hall *et al.*, 1975; Hall, 1977 and Dollard, 1979).

3.2.2. Root uptake of trace metals

This topic probably accounts for the greater part of the huge literature on plants and trace metals. Chapters in the first volume of this book should be consulted for the physiology of root uptake of particular trace metals. Hughes *et al.* (1980) have recently reviewed this literature in an ecological context. The three main areas covered were: soil factors affecting 'availability' of trace metals to roots, movement of trace metals to roots and absorption of trace metals by roots. Commenting on a tabulation of results of many studies of trace metal absorption by whole plants (usually crops), Hughes *et al.* (1980) conclude that 'the main factors that influence whole plant uptake are soil pH and the presence and levels of other ionic species'. They also observed considerable inter- and intra-specific differences in trace metal uptake (Jarvis *et al.*, 1976; John and van Laerhoven, 1976).

3.2.3. Movement of trace metals in plants

Within-season changes in trace metal concentrations of particular plant parts have been reported by Guha and Mitchell (1966), Tyler (1971), Hall (1977) and Dollard (1979). Some of these changes may be due to pluvial or leaching losses, but others cannot be. The mechanisms of trace metal movement within plants are little understood. It seems likely that most metals become chelated by relatively simple agents in xylem sap (Tiffin, 1972, 1977; Lyon *et al.*, 1969). Some, particularly cadmium, are probably not chelated in this way (Petit and van de Geijn, 1978; van de Geijn and Petit, 1978). Mechanisms of immobilization exist (Malone *et al.*, 1974) and fixation of trace metals in xylem walls may occur (Hall, 1977; Dollard, 1979; Tan, 1980).

3.2.4. Localization of trace metals in soil

There are many reports of decreasing trace metal concentrations with increasing depth below soil surface (Fortescue, 1974; Heinrichs and Mayer, 1977; Hutchinson and Whitby, 1974; John *et al.*, 1976*a*,*b*; Ragaini *et al.*, 1977; Smith, 1976; Swaine and Mitchell, 1960; Tyler, 1972). It has been suggested (Tyler, 1972) that this pattern of trace metal distribution is causally related to the rather similar pattern of organic matter distribution in soil profiles. As the cuticle of shed leaves breaks down in the early stages of decomposition, cell walls are exposed providing sites for passive sorption (Rühling and Tyler, 1971; Tyler, 1972; Roberts, 1975). Consequently, recently fallen litter has a higher concentration of trace metals than living leaves (Nilsson, 1972). As decomposition proceeds, trace metal concentrations increase (Tyler, 1971, 1972; Palmer, 1972; Martin and Coughtrey, 1975; Lawrey, 1978; Denaeyer-De Smet, 1974). Thus litter and soil organic matter represent reservoirs of organic exchange sites on which cations may accumulate. However, not all observations of trace metal movement and localization in soil can be accounted for by this appealingly simple model. In particular, chelation is of importance for those metals which tend to form stable complexes, such as copper and lead (Begovich and Jackson, 1975). This is confirmed by the enhancement of trace metal mobility in soil by the presence of soluble organic compounds (Witkamp and Ausmus, 1976; Lossaint, 1959; Baker, 1973; Bolter and Butz, 1975). This effect is particularly marked for those trace metals which readily form complexes. Just as mobile (i.e. dissolved) organic compounds increase the soil mobility of readily complexed trace metals, so fixed (i.e. non-dissolved) organic compounds, in litter and humus, provide reservoirs of complexing agents most particularly for the same trace metals. Consequently, soils containing very high amounts of organic matter (e.g. the mor of coniferous forest) show least leaching from litter for the most readily complexed trace metals such as lead and copper, and most leaching for less readily complexed metals such as zinc and cadmium (Tyler, 1978). Metal mobility is also related to sorption by clays and hydrous metal oxides (Ellis and Knezek, 1972; Page, 1974). Trace metals may also be rendered immobile by the formation of sulphides (this is redox dependent), carbonates or phosphates, all of which depend on a number of facets of soil chemistry, not notably pH. Clearly the location, mobility and chemical form of soil trace metals determine their availability to plants for recycling. Furthermore, the more mobile a trace metal, the more readily it may be lost from the ecosystem to ground water or stream flow as a result of some vertical or lateral water movement.

3.2.5. Trace metals in food chains

The field evidence for food chain transport of certain trace metals has recently been reviewed by Hughes *et al.* (1980). After searching a very wide literature they found relatively few reports based on the systematic sampling of consumer populations and their food on a comparable basis.

However, they were able to make certain generalizations, notably the following:

(a) In both vertebrates and invertebrates, lead concentrations in animals tend to increase as the lead concentration of their food increases. They point out that some of this lead burden may be derived directly from the atmosphere. There is lead contamination of animals at different trophic levels in ecosystems but there is little evidence of food-chain concentration of lead, except in some carnivorous invertebrates, and in earthworms under certain conditions. However, lead is not evenly distributed within animals, being particularly associated with calcified tissues in both vertebrates and invertebrates. In addition, certain organs such as the kidneys or the liver tend to be 'target organs' with unusually high lead concentrations in vertebrates.

(b) Cadmium has great mobility in ecosystems. Food-chain concentration of cadmium appears to take place. This may be explained in part by the lack of any tendency to associate with calcified tissues and hence a greater availability of cadmium to consumers by virtue of being held in easily digested soft tissues. As with lead, marked differences in cadmium concentration are observed between organs.

(c) In addition to the known ability of vertebrates to regulate body zinc concentrations, the field data reviewed by Hughes *et al.* (1980) suggest that at least some invertebrates possess this capability. In general, the concentration of zinc in animals varies very much less than in their food.

The extensive literature on mercury contains evidence of food-chain concentration, particularly from graminivores to carnivores where mercury-containing seed dressing has been used (Vostal, 1972). Johnels *et al.* (1979) warn that such transfers may acquire a new significance as alkyl mercury seed dressings are replaced by industry and fossil fuel combustion as the principal sources of mercury in the terrestrial environment. Transfer of copper in food chains has been studied principally for those detritivores possessing the copper-containing respiratory pigment haemocyanin. These are the isopods (Wieser *et al.*, 1976) and the pulmonate molluscs (Coughtrey and Martin, 1976, 1977; Moser and Wieser, 1979). The copper content of these organisms is often related to, but higher than, that in their food. In a study of a relatively uncontaminated Northern Great Plains site (Munshower and Neuman, 1979), where plant-available copper and zinc appeared to be marginal, concentrations of both metals increased from soil to plants to animal soft tissues—the animals were ungulates, a lagomorph and game birds. This type of apparent food-chain concentration, where consumer metal concentrations are constrained to a narrow range by physiological requirements (see discussion of zinc above) should be distinguished from the apparently uncontrolled influx to consumers shown by cadmium.

Martin and Coughtrey (Chapter 4) discuss food-chain transfer of trace metals in the context of their ecological impact. Their chapter should be referred to for evidence of transfer of a number of elements (e.g. selenium, thallium) not covered by Hughes *et al.* (1980).

4. ECOSYSTEM AUDITS AND WATERSHED STUDIES

As explained in Section 2.1 of this chapter, ecosystem audits involve the quantification of the distribution of an element within an ecosystem and the estimation of transfers between the ecosystem's major components. The results presented in this book by Martin and Coughtrey (Chapter 4) are a detailed example of this. Comparisons between ecosystem audits made by different authors are difficult, not least because different sampling strategies and analytical techniques are often used. Few authors divide up the biotic and soil components of the ecosystem in the same way and similarly, few choose to analyse for the same elements. Problems of methodology are particularly acute in the case of soil analyses, where techniques of preparation and extraction vary not only between authors but between the same authors in subsequent papers. Consequently, it is only possible to draw very general conclusions on the basis of comparing ecosystem audits of trace metals at the present. In Table 3.1 summary ecosystem audits for lead in nine ecosystems are presented. Whilst atmospheric inputs are only reported for four ecosystems, they have all been arranged in what is thought to be an order of contamination from left to right. Given the wide range of biomasses in these ecosystems it is of note that the quantity of lead in the vegetation falls within the narrow range 38–198 mg/m^2 in all but the grossly polluted woodland near an English smelter (9). The quantity of lead in litter ranges from 57 mg/m^2 in the dry heath (3) to 480 mg/m^2 in the Spruce forest (5), broadly following the quantity of litter found in the ecosystem. Only the contaminated woodland (9) lies outside this range. Soil lead estimates are very much greater, being of the order of one thousand mg/m^2 for extractable lead in uncontaminated regions and in similar situations of the

TABLE 3.1
ECOSYSTEM AUDITS FOR LEAD

Ecosystem	(1) Shore meadow	(2) Wet heath	(3) Dry heath	(4) Mixed deciduous forest	(5) Spruce forest	(6) Beech forest	(7) Urban savannah	(8) Urban wetland	(9) Deciduous woodland
Stocks (*mg*/*m*²)									
Tree canopy	—	—	—	32·4	65·4	46·2 }	98	—	1 424·5
Stem	—	—	—	40·8	18·8	32·7 }		—	990·2
Lower strata	1·24	19	21	—	—	—	3·1	5·3	823·3
Roots	165·7	29	17	54·9	78	38·2	96·8	49·4	?
Vegetation total	166·94	48	38	128·1	162·2	117·1	197·9	54·7	3 238 +
Litter	14·3	111	57	342	480·4	419·7	440	140	30 598
Litter and vegetation	181·24	159	95	470·1	642·6	536·8	637·9	194·7	33 836 +
Soil 'extractable'	1 202	1 330	1 350	?	?	?	?	?	?
Soil 'total'	?	?	?	64 000	10 000	6 330	18 300	18 400	48 327
Transfers (*mg*/*m*²/*yr*)									
Atmospheric input	?	?	?	28·6	?	?	81·5	81·5	285·3
Retention by biomass	?	?	?	1·6	20·9	3·6	?	?	?
Litter fall, etc.[a]	?	?	?	12·5	10·3	10·8	1·04	0·12	294·7
Output in ground water or streamflow	?	?	?	0·16	?	?	1·1	1·1	?
Author(s)	Tyler, 1971	Tyler *et al.*, 1973	Tyler *et al.*, 1973	van Hook *et al.*, 1977	Denaeyer-De Smet, 1974	Denaeyer-De Smet, 1974	Parker *et al.*, 1978	Parker *et al.* 1978	Martin and Coughtrey, Chapter 4

[a] This includes all litter plus leachate and throughfall where given.
A dash (—) indicates no measurement possible.
A question mark (?) indicates no measurement available.

order of tens of thousands of mg/m^2 for total lead. A broad pattern emerges of partition between compartments thus:

Total soil lead, $n \times 10^4$ mg/m^2; extractable soil lead, $n \times 10^3$ mg/m^2; litter lead, $n \times 10^1$–10^2 mg/m^2; plant lead, $n \times 10^1$–10^2 mg/m^2.

Soil lead is the major pool of the element even in relatively uncontaminated ecosystems. In the case of the grossly polluted woodland (9) all values are very mugh higher, the relative importance of plant lead (underestimated through lack of a root estimate) and litter lead being increased relative to total soil lead. The plant lead level was more than sixteen times greater than in the next highest case (7), whilst the litter lead was almost sixty-four times greater than in the next highest (5).

Broadly similar patterns have been found elsewhere. Siccama and Smith (1978), working in a New England hardwood forest remote from pollution and with an atmospheric input very similar to example (4) in Table 3.1, estimated plant lead at either 70 or 100 mg/m^2 according to the chemical analysis used, and forest floor lead at 1460 mg/m^2. Atmospheric inputs, even at the polluted site (9) are very small compared to soil and litter lead pools. Losses in stream flow or by leaching are even smaller. In addition to the figures given in Table 3.1, this great excess of lead input over output is reported for other ecosystems by Heinrichs and Mayer (1977), Bowen (1975) and Rolfe and Haney (1975).

Table 3.2 gives summary ecosystem audits for copper. Interestingly, the highest plant copper level is in the shore meadow (1). This is almost entirely associated with roots, which may be live or dead, and may be contaminated with soil to some extent, in spite of the care taken in analysis. Thus plant copper in the heavily polluted woodland (9) is not greatly different to that in some less- or non-polluted ecosystems. Where data are available, litter copper is rather smaller than plant copper and is broadly proportional to the amount of litter present. Litter copper at site (9) is very much higher than in the other ecosystems (more than one hundred fold) and total soil copper is about twenty times greater than in the other sites. The relationship of copper input to output is only reported for two of the ecosystems in Table 3.2, an amount equal to over one-third of input copper being lost in ground water. Heinrichs and Mayer (1977) found outputs close to one-half of input copper in spruce and beech forests. In calculations for the Upper Thames Basin, Bowen (1975) found a ratio between copper input and drainage output of 11·84:1, although inclusion of loss by cropping increased the proportion output to over half.

Table 3.3 gives summary ecosystem audits for cadmium. Roots appear

TABLE 3.2

ECOSYSTEM AUDITS FOR COPPER

Ecosystem	(1) Shore meadow	(2) Wet heath	(3) Dry heath	(4) Northern hardwood forest	(5) Spruce forest	(6) Beech forest	(7) Urban savannah	(8) Urban wetland	(9) Deciduous woodland
Stocks (mg/m^2)									
Tree canopy	—	—	—	21·5	30·3	74·4 }	26·4 }	—	76·7
Stem	—	—	—	20·2	65·7	37·9 }		—	124·5
Lower strata	1·46	6·7	5·7	0·24	—	—	1·4	1·4	14·9
Roots	298·6	14·1	8·2	20	18·0	23·5	29·7	33·5	?
Vegetation total	300·06	20·8	13·9	61·94	114	135·8	57·5	34·9	216·1 +
Litter	5·7	22·2	11·1	32	?	?	80	20	1 590
Litter and vegetation	305·76	43	25	93·97	?	?	137·5	54·9	1 806·1 +
Soil 'extractable'	1 715·6	1 150	690	65	?	?	?	?	?
Soil 'total'	?	?	?	?	?	?	450	550	9 265
Transfers ($mg/m^2/yr$)									
Atmospheric input	?	?	?	?	?	?	16·4	16·4	26·3
Retention by biomass	?	?	?	1·22	9·9	4·1	?	?	?
Litter fall, etc.	?	?	?	5·41	2·1	9·0	0·2	0·02	31·5
Output in ground water or streamflow	?	?	?	?	?	?	6·5	6·5	?
Author(s)	Tyler, 1971	Tyler *et al.*, 1973	Tyler *et al.*, 1973	Whittaker *et al.*, 1979	Denaeyer-De Smet, 1973	Denaeyer-De Smet, 1973	Parker *et al.*, 1978	Parker *et al.*, 1978	Martin and Coughtrey, Chapter 4

A dash (—) indicates no measurement possible.
A question mark (?) indicates no measurement available.

TABLE 3.3

ECOSYSTEM AUDITS FOR CADMIUM

Ecosystem	(1) Wet heath	(2) Dry heath	(3) Mixed deciduous forest	(4) Urban savannah	(5) Urban wetland	(6) Deciduous woodland
Stocks (mg/m^2)						
Tree canopy	—	—	2·1	1·3	—	32·4
Stem	—	—	4·4		—	29·2
Lower strata	0·1	0·1	—	0·11	0·05	3·6
Roots	1·2	0·6	2·8	2·19	8·7	?
Vegetation total	1·3	0·7	9·3	3·6	8·75	65·2+
Litter	0·9	0·4	5·2	5	1·3	531·6
Litter and vegetation	2·2	1·1	14·5	8·6	10·05	596·8+
Soil 'extractable'	26	28	?	?	?	?
Soil 'total'	?	?	300	476	446	4 348
Transfers ($mg/m^2/yr$)						
Atmospheric input	?	?	2·1	0·82	0·82	9·2
Retention by biomass	?	?	0·12	?	?	?
Litter fall, etc.	?	?	0·59	0·02	0·002	11·2
Output in ground water or streamflow	?	?	0·7	0·12	0·12	?
Author(s)	Tyler *et al.*, 1973	Tyler *et al.*, 1973	van Hook *et al.*, 1977	Parker *et al.*, 1978	Parker *et al.*, 1978	Martin and Coughtrey, Chapter 4

A dash (—) indicates no measurement possible.
A question mark (?) indicates no measurement available.

to be of particular importance as a cadmium reservoir, although the role of dead roots and soil contamination in this is not clear. Levels in litter are broadly similar in scale to those in plants, the role of litter being superproportionally increased at the most polluted site (6). This role of litter and organic matter as a cadmium sink has been noted by Somers (1978). An amount equivalent to two-thirds of atmospheric input appears to be retained by the mixed deciduous forest (3) as much as six-sevenths by the urban savannah and wetland ((4) and (5)). Only one-third of input was retained by the spruce and beech forests investigated by Heinrichs and Mayer (1977), whilst the Upper Thames Basin retained 54 % of inputs from atmosphere and fertilizer, 36 % after removal in crops (Bowen, 1975).

Table 3.4 gives summary ecosystem audits for zinc. Again, roots are an important pool and overall plant zinc relates to biomass. Plant zinc levels range from 71 mg/m^2 to 825 mg/m^2, excepting the most polluted site (10) with 3120 mg/m^2. In both the shore meadow (1) and urban wetlands (9) almost all plant zinc is in the roots. Litter zinc ranges from 18·3 mg/m^2 to 781 mg/m^2, most being below 100 mg/m^2. Litter zinc is often less than plant zinc. The main exception is the very contaminated site (10), where a superproportional storage in litter seems to occur. In all systems the soil is the main zinc pool, soil exchangeable zinc levels being ten to twenty times litter zinc, total soil zinc levels one hundred to a thousand times greater. In the systems for which data are given in Table 3.4 ((8) and (9)) 85 % of input zinc is retained. Beech and spruce forests retained 72 % and 51 %, respectively (Heinrichs and Mayer, 1977). Bowen (1975) calculated a retention of 98 % after drainage for the Upper Thames Basin, although cropping losses resulted in a net outflow.

In the case of each of these four elements, Pb, Cu, Zn and Cd, atmospheric input is very small in comparison to the soil pool. As this input is directly to the vegetation and thence to the litter, its size relative to these pools is, however, more important.

Even within the vegetation the distribution of trace metals is far from uniform. In terms of concentrations, particularly of lead and zinc, bark, twigs and leaves are important. High concentrations of trace metals in twigs may well be due in large part to the high proportion of bark in twigs. High concentrations of trace metals in bark have been reported by Denaeyer-De Smet (1974) for *Picea*, by Hall (1977) for *Ligustrum*, *Crataegus* and *Acer*, by Lemmey (1976) for *Larix* and Martin and Coughtrey (Chapter 4) for *Quercus*. Denaeyer-De Smet (1974) did not find this effect in *Fagus*. There are many other records of tissue concentrations of trace metals in woody plants, a number of which were summarized by Hall *et al.* (1975). Lead

TABLE 3.4

ECOSYSTEM AUDITS FOR ZINC

Ecosystem	(1) Shore meadow	(2) Wet heath	(3) Dry heath	(4) Northern hardwood forest	(5) Mixed deciduous forest	(6) Spruce forest	(7) Beech forest	(8) Urban savannah	(9) Urban wetland	(10) Deciduous woodland
Stocks (mg/m^2)										
Tree canopy	—	—	—	293·3	80·2	223·1	252·8	414	—	1 654
Stem	—	—	—	197·8	152·0	279·3	132·9		—	1 265·5
Lower strata	5·9	35	28	1·24	—	—	—	11	8	249·5
Roots	209·2	95	43	204	177·7	76·8	56	388	817·4	?
Vegetation total	215·1	130	71	696·34	409·9	579·2	441·7	813	825·4	3 169·0+
Litter	18·3	94	46	300	781	?	?	100	33	27 107
Litter and vegetation	233·4	224	117	996·34	1 190·9	?	?	913	858·4	30 226·7+
Soil 'extractable'	784·4	1 720	1 730	470	?	?	?	?	?	?
Soil 'total'	?	?	?	?	141 700	?	?	107 100	123 100	255 500
Transfers ($mg/m^2/yr$)										
Atmospheric input	?	?	?	?	?	?	?	98·4	98·4	599·9
Retention by biomass	?	?	?	14·9	?	?	?	?	?	?
Litter fall, etc.	?	?	?	80·3	?	?	?	2·52	0·29	681·3
Output in ground water or streamflow	?	?	?	?	?	?	?	15	15	?
Author(s)	Tyler, 1971	Tyler *et al.*, 1973	Tyler *et al.*, 1973	Whittaker *et al.*, 1979	van Hook *et al.*, 1977	Denaeyer-De Smet, 1973	Denaeyer-De Smet, 1973	Parker *et al.*, 1978	Parker *et al.*, 1978	Martin and Coughtrey, Chapter 4

A dash (—) indicates no measurement possible.
A question mark (?) indicates no measurement available.

concentrations were generally higher in bark than wood or leaves, the enhancing effect of pollution being relatively greater in bark. The pattern was broadly similar for zinc, but did not appear to apply to copper. Thus bark forms a secondary perennial focus of lead and zinc concentration in forest ecosystems after the forest floor. Although it does not constitute a large elemental pool, due to its relatively small biomass, it does occupy an important position in the ecosystem. This arises both from its closeness to the vascular cambium and its status as the carrier of stem flow from canopy to soil.

5. LOCI OF TRACE METAL CONCENTRATION IN ECOSYSTEMS

It is clear that enhanced concentrations of some trace metals may arise in ecosystem components by virtue of retention exceeding loss. This is the case in litter and bark, in some other plant parts according to species, in bryophytes (see Puckett and Burton, Chapter 7) and in some animals, particularly carnivores. The particular patterns of accumulation vary according to the trace metal involved. It is implicit in most discussions of these concentrations that the mean values cited in each case are representative of the burden in that biotic component. This may well be untrue. Frequency distributions of trace metal concentrations in individuals or replicate tissue samples may be markedly skewed (Beardsley *et al.*, 1978; Garten *et al.*, 1977; Gish and Christensen, 1973; Wallace *et al.*, 1977). Garten *et al.* (1977) made the interesting observation that concentrations of essential elements were less skewed than those of non-essential elements (these were positively skewed). Positive skewing of the population frequency distribution of toxic trace metal concentrations will result in sublethal or lethal effects occurring before the mean concentration reaches a toxic level. This could have important consequences for the interpretation of data on trace metal accumulations in ecosystems.

6. CONCLUSIONS

(a) The atmospheric input of trace metals to ecosystems has increased in the last two centuries, not only in areas close to pollution sources but also in more remote areas.

(b) The trace metals each have characteristic modes of uptake, retention

and loss by plants and animals. In food chains, some concentration of lead may take place in carnivores, the element tending to be most in evidence in calcareous tissues. Cadmium is extraordinarily mobile, showing increasing concentrations in herbivores then carnivores in many cases. It is located principally in the soft, easily digested tissues. Zinc uptake, retention and loss appears to be regulated in a large number of animal species, including invertebrates.

(c) Soil is the largest pool of lead, copper, cadmium and zinc in those ecosystems for which data are available. Litter, whether 'forest floor' or 'attached dead' is usually the next largest pool, except for copper where vegetation is often more important. Lead, zinc and cadmium concentrations in the forest floor and in soil are often related to organic matter content. Whilst bark does not constitute a major pool for trace metals, it is often a focus of high concentrations of lead, zinc and cadmium, but not so markedly of copper.

(d) Whilst atmospheric inputs of trace metals are small, they are very largely retained by ecosystems, particularly in the case of lead. Much of the retained trace metal input is concentrated in the top few centimetres of the soil, a region of immense importance for ecosystem function. Such effects have been observed at sites remote from urban and industrial activity.

(e) The pattern of differential mobility of trace elements in various ecosystem components, combined with skewed frequency distributions in trace metal concentrations, suggests that toxic effects associated with concentration might well arise in individual components of ecosystems whose mean concentration of the element might not imply this. To detect the occurrence of such events and assess their significance represents a considerable scientific challenge.

REFERENCES

Baker, W. E. (1973). The role of humic acids from Tasmanian podzolic soils in mineral degradation and metal mobilization. *Geochim. Cosmochim. Acta*. **37**: 269–81.

Beardsley, A., M. J. Vagg, P. H. T. Beckett and B. F. Sansom (1978). Use of the field vole (*M. agrestis*) for monitoring potentially harmful elements in the environment. *Environ. Pollut*. **16**: 65–71.

Begovich, C. L. and D. R. Jackson (1975). Documentation and application of SCHEM. ORNL-NSF-EATC, Oak Ridge National Laboratory, Tennessee, pp. 1–49.

Biggins, P. D. E. and R. M. Harrison (1978). Identification of lead compounds in urban air. *Nature, Lond*. **272**: 531–2.

Boggess, W. R. and B. G. Wixson (1977). Lead in the environment. Report PB-278-278, National Technical Information Service (USA). 272 pp.

Bolter, E. and T. R. Butz (1975). Heavy metal mobilization by natural organic acids. In: *Symposium Proceedings, Vol. II, International Çonference on Heavy Metals in the Environment* (T. C. Hutchinson (ed)), Toronto, Canada, pp. 353–62.

Bormann, F. H. and G. E. Likens (1979). *Pattern and Process in a Forested Ecosystem*. Springer-Verlag, New York. 253 pp.

Bowen, H. J. M. (1975). Residence times of heavy metals in the environment. In: *Symposium Proceedings, Vol. I, International Conference on Heavy Metals in the Environment* (T. C. Hutchinson (ed)), Toronto, Canada, pp. 1–19.

Brown, A. H. F. (1974). Nutrient cycles in oakwood ecosystems in N. W. England. In: *The British Oak* (M. G. Morris and F. H. Perring, B.S.B.I. (eds)), pp. 141–61.

Chamberlain, A. (1967). Deposition of particles to natural surfaces. Airborne Microbes. *Symp. Soc. Gen. Microbiol.* **17**: 138–64.

Chamberlain, A. C., M. J. Heard, P. Little, D. Newton, A. C. Wells and R. D. Wiffen (1978). Investigations into lead from motor vehicles. UKAEA, Harwell, 151 pp.

Coughtrey, P. J. and M. H. Martin (1976). The distribution of Pb, Zn, Cd and Cu within the pulmonate mollusc *Helix aspersa* Müller, *Oecologia, Berl.* **23**: 315–22.

Coughtrey, P. J. and M. H. Martin (1977). The uptake of lead, zinc, cadmium and copper by the pulmonate mollusc *Helix aspersa* Müller, and its relevance to the monitoring of heavy metal contamination of the environment. *Oecologia, Berl.* **27**: 65–74.

Daines, R. H., H. Motto and D. M. Chilko (1970). Atmospheric lead: Its relationship to traffic volume and proximity to highways. *Environ. Sci. Technol.* **4**: 318–22.

Denaeyer-De Smet, S. (1973). Comparaison du cycle biologique annuel de divers oligoéléments dans une pessière (Piceetum) at dans une hêtraie (Fagetum) éstablies sur même roche-mère. *Bull. Soc. Roy. Bot. Belg.* **106**: 149–65.

Denaeyer-De Smet, S. (1974). Cycle biologique annuel et distribution du plomb dans une pessière (Piceetum) et une hêtraie (Fagetum) éstablies sur même roche-mère. *Bull. Soc. Roy. Bot. Belg.* **107**: 115–25.

Dollard, G. J. (1979). Some aspects of the behaviour of heavy metal ions in the tissue of a woody plant. Ph.D. Thesis, Liverpool Polytechnic (Council for National Academic Awards), Liverpool. 277 pp.

Elias, R., Y. Hirao and C. Patterson (1975). Impact of present levels of aerosol Pb concentrations on both natural ecosystems and humans. In: *Symposium Proceedings, Vol. II, International Conference on Heavy Metals in the Environment* (T. C. Hutchinson (ed)), Toronto, Canada, pp. 251–7.

Ellis, B. G. and B. D. Knezek (1972). Adsorption reaction of micronutrients soils. In: *Micronutrients in Agriculture* (J. J. Mortvedt, P. M. Giordano and W. L. Lindsay (eds)), Soil Sci. Soc. America, Madison, Wisconsin, pp. 59–78.

Fortescue, J. A. C. (1974). The environment and landscape geochemistry. *Western Miner*. March 1974.

Garten, C. T., J. B. Gentry and R. R. Sharitz (1977). An analysis of elemental concentrations in vegetation bordering a south eastern United States coastal plain stream. *Ecology*. **58**: 979–92.

Gish, C. D. and R. E. Christensen (1973). Cadmium, nickel, lead and zinc in earthworms from roadside soil. *Environ. Sci. Technol.* **7**: 1060–2.

Grace, J. (1977). *Plant Responses to Wind*. Academic Press, London. 204 pp.

Guha, M. M. and R. L. Mitchell (1966). Trace and major element composition of the leaves of some deciduous trees. *Plant Soil*. **24**: 90–112.

Haghiri, F. (1974). Plant uptake of cadmium influenced by cation exchange capacity, organic matter, zinc and soil temperature. *J. Environ. Qual.* **3**: 180–3.

Hall, C. (1977). A study of certain heavy metals in woody plants. Ph.D. Thesis, Liverpool Polytechnic (Council for National Academic Awards), Liverpool. 208 pp.

Hall, C., M. K. Hughes, N. W. Lepp and G. J. Dollard (1975). Cycling of heavy metals in woodland ecosystems. In: *Symposium Proceedings, Vol. II, International Conference on Heavy Metals in the Environment* (T. C. Hutchinson (ed)), Toronto, Canada, pp. 257–71.

Harrison, S. J., N. W. Lepp and D. A. Phipps (1978). Uptake of copper by excised roots. I. A modified experimental technique for measuring ion uptake by excised roots and its application in determining uptake characteristics of 'free' copper ions in excised *Hordeum* roots. *Z. Pflanzenphysiol.* **90**: 443–50.

Heinrichs, H. and R. Mayer (1977). Distribution and cycling of major and trace elements in two Central European forest ecosystems. *J. Environ. Qual.* **6**: 402–7.

Hemphill, D. D. and J. H. Rule (1975). Translocation of ^{210}Pb and ^{109}Cd by plants. In: *Symposium Proceedings, Vol. II, International Conference on Heavy Metals in the Environment* (T.C. Hutchinson (ed)), Toronto, Canada, pp. 72–86.

Horler, D. N. H. and J. Barber (1979). Relationships between vegetation and heavy metals in the atmosphere. In: *Int. Conf. Management and Control of Heavy Metals in the Environment* (R. Perry (chairman)), C.E.P. Consultants, Edinburgh, pp. 275–8.

Hughes, M. K., N. W. Lepp and D. A. Phipps (1980). Aerial heavy metal pollution and terrestrial ecosystems. *Adv. Ecol. Res.* **11**: 217–327.

Hutchinson, T. C. and L. M. Whitby (1974). Heavy metal pollution in the Sudbury mining and smelting region of Canada. I. Soil and vegetation contamination by nickel, copper and other metals. *Environ. Conserv.* **1**: 123–32.

Jarvis, S. C., L. H. P. Jones and M. J. Hopper (1976). Cadmium uptake from solution by plants and its transport from roots to shoot. *Plant Soil*. **44**: 179–91.

John, M. K. and G. J. van Laerhoven (1976). Differential effects of cadmium on lettuce varieties. *Environ. Pollut.* **10**: 163–73.

John, M. K., C. van Laerhoven and C. Cross (1976*a*). Cadmium, lead and zinc accumulation in soils near a smelter complex. *Environ. Lett.* **10**: 23–35.

John, M. K., C. van Laerhoven and J. Bjerring (1976*b*). Effect of a smelter complex on the regional distribution of cadmium, lead and zinc in litters and soil horizons. *Arch. Env. Contam.* **4**: 456–68.

Johnels, A., G. Tyler and T. Westermark (1979). A history of mercury levels in Swedish fauna. *Ambio*. **8**: 160–8.

Jordan, M. J. (1975). Effects of zinc smelter emissions and fire on a chestnut–oak woodland. *Ecology*. **56**: 78–91.

Lawrey, J. D. (1978). Trace metal dynamics in decomposing leaf litter in habitats variously influenced by coal strip mining. *Can. J. Bot*. **56**: 953–62.

Lee, R. E., Jr. and D. J. von Lehmden (1973). Trace metal pollution in the environment. *J. Air Pollut. Control Assoc*. **23**: 853–7.

Lemmey, R. D. (1976). Heavy metals in Japanese larch plantations. Unpublished B.Sc. Dissertation, Liverpool Polytechnic, Liverpool. 46 pp.

Likens, G. E., F. H. Bormann, R. S. Pierce, J. S. Eaton and N. M. Johnson (1977). *Biogeochemistry of a Forested Ecosystem*. Springer-Verlag, New York. 146 pp.

Little, P. and R. D. Wiffen (1977). Emission and deposition of petrol exhaust Pb—I. Deposition of exhaust Pb to plant and soil surfaces. *Atmos. Env*. **11**: 437–47.

Livett, E. A., J. A. Lee and J. H. Tallis (1979). Lead, zinc and copper analyses of British blanket peats. *J. Ecol*. **67**: 865–92.

Lossaint, P. (1959). Etude expérimentale de la mobilisation du fer des sols sous l'influence des litières forestières. *Ann. Agron*. **10**: 369–542.

Lundgren, D. A. (1971). Determination of particulate composition, concentration and size distribution changes with time. *Atmos. Env*. **5**: 645–65.

Lyon, G. L., P. J. Peterson and R. R. Brooks (1969). Chromium-51 distribution in tissues and extracts of *Leptospermum scoparium*. *Planta*. **88**: 282–7.

Malone, C., D. E. Koeppe and R. J. Miller (1974). Localization of lead accumulated by corn plants. *Plant Physiol*. **53**: 388–94.

Martin, M. H. and P. J. Coughtrey (1975). Preliminary observations on the levels of cadmium in a contaminated environment. *Chemosphere*. **3**: 155–60.

Melton, C. W., R. I. Mitchell, D. A. Trayser and J. F. Foster (1973). Chemical and physical characteristics of automotive exhaust particulate matter in the atmosphere. NTIS, Report No. EPA-650/2 73 001.

Moran, J. B., M. J. Baldwin, O. J. Manary and J. C. Valenta (1972). Effects of fuel additives on the chemical and physical characteristics of particulate emissions in automobile exhaust. US Environmental Protection Agency, PB-222, 799.

Moser, H. and W. Wieser (1979). Copper and nutrition in *Helix pomatia* (L). *Oecologia, Berl*. **42**: 241–51.

Munshower, F. F. and D. R. Neuman (1979). Pathways and distribution of some heavy metals in a grassland ecosystem. In: *Int. Conf. Management and Control of Heavy Metals in the Environment* (R. Perry (chairman)), C.E.P. Consultants, Edinburgh, pp. 206–9.

Nilsson, I. (1972). Accumulation of metals in spruce and needle litter. *Oikos*. **23**: 132–6.

Nriagu, J. O. (1979). Global inventory of natural and anthropogenic emissions of trace metals to the atmosphere. *Nature, Lond*. **279**: 409–11.

Page, A. L. (1974). Fate and effects of trace elements in sewage sludge when applied to agricultural lands. A literature review study. US Environmental Protection Agency, Report No. EPA-670/2-74-005. 108 pp.

Palmer, K. T. (1972). Lead uptake in sycamore (*Platanus occidentalis* L.). Ph.D. Thesis, University of Missouri, Columbia. 144 pp.

Parker, G. R., W. W. McFee and J. M. Kelly (1978). Metal distribution in forested ecosystems in urban and rural northwestern Indiana. *J. Environ. Qual*. **7**: 337–42.

Petit, C. M. and S. C. van de Geijn (1978). *In vivo* measurement of cadmium (^{115}Cd), transport and accumulation in stems of intact tomato plants (*Lycopersicon esculentum* Mill). 1. Long distance transport and accumulation. *Planta*. **138**: 137–43.

Quarles, H. D., R. B. Hanawalt and W. E. Odum (1974). Lead in small mammals, plants and soil at varying distances from a highway. *J. Appl. Ecol.* **11**: 937–49.

Ragaini, R. C., H. R. Ralston and N. Roberts (1977). Environmental trace metal contamination in Kellog, Idaho, near a lead smelting complex. *Environ. Sci. Technol.* **11**: 773–81.

Randall, C. W., T. J. Grizzard, R. C. Hoehn and D. R. Helsel (1979). The origin, distribution and fate of heavy metals in storm water run-off. In: *Int. Conf. Management and Control of Heavy Metals in the Environment* (R. Perry (chairman)), C.E.P. Consultants, Edinburgh, pp. 239–42.

Roberts, T. M. (1972). Plants as monitors of airborne metal pollution. *J. Env. Plan. Pollut. Cont.* **1**: 43–54.

Roberts, T. M. (1975). A review of some biological effects of lead emissions from primary and secondary smelters. In: *Symposium Proceedings, Vol II, International Conference on Heavy Metals in the Environment* (T.C. Hutchinson (ed)), Toronto, Canada, pp. 503–32.

Robinson, E. and F. L. Ludwig (1964). Size distribution of sulfur-containing compounds in urban aerosols. *J. Coll. Sci.* **20**: 571–84.

Rolfe, G. L. and A. Haney (1975). *An Ecosystem Analysis of Environmental Contamination by Lead.* Institute for Environmental Studies, University of Illinois at Urbana–Champaign, 133 pp.

Rühling, Å. and G. Tyler (1968). An ecological approach to the lead problem. *Bot. Notisk.* **121**: 321–42.

Rühling, Å. and G. Tyler (1969). Ecology of heavy metals—A regional and historical study. *Bot Notisker.* **122**: 248–59.

Rühling, Å. and G. Tyler (1971). Anrikning av tungmetalleri barrskog vid Finspång, Trollhättan, Oskarshamn och Fliseryd. Rapport No. 23 från Ekologiska tungmetallundersökningar (Lund Mimeographed). 14 pp.

Semb, A. (1978). Deposition of trace elements from the atmosphere in Norway. SNSF Project FR.13/78, Ås, Norway. 28 pp.

Siccama, T. G. and W. H. Smith (1978). Lead accumulation in a northern hardwood forest. *Env. Sci. Technol.* **12**: 593–4.

Smith, W. H. (1974). Air pollution—Effects on the structure and function of the temperate forest ecosystem. *Environ. Pollut.* **6**: 111–29.

Smith, W. H. (1976). Lead contamination of the roadside ecosystem. *J. Air. Poll. Cont. Assoc.* **26**: 753–66.

Somers, G. F. (1978). The role of plant residues in the retention of cadmium in ecosystems. *Environ. Pollut.* **17**: 287–95.

Swaine, D. J. and R. L. Mitchell (1960). Trace element distribution in soil profiles. *J. Soil Sci.* **11**: 347–68.

Tan, T. K. (1980). The physiological basis for producing lead pollution histories from tree-rings. Ph.D. Thesis, Liverpool Polytechnic (Council for National Academic Awards), Liverpool. 283 pp.

Tiffin, L. O. (1972). Translocation of micronutrients in plants. In: *Micronutrients in Agriculture* (J. J. Mortvedt, P. M. Giordano and W. L. Lindsay (eds)), Soil Sci. Soc. America, Madison, Wisconsin, pp. 199–229.

Tiffin, L. O. (1977). The form and distribution of metals in plants. An overview. In *Biological Implications of Metals in the Environment*, Proc. 15th Ann. Hanford Life Sci. Symp. ERDA-TIC-CONF, No. 750929, Oak Ridge, Tennessee, pp. 315–34.

Tukey, H. B., Jr (1971). Leaching of substances from plants. In: *Ecology of Leaf Surface Microorganisms* (T. F. Preece and C. H. Dickinson (eds)). Academic Press, London, pp. 67–80.

Tukey, H. B., H. B. Tukey, Jr and S. H. Wittwer (1958). Loss of nutrients by foliar leaching, as determined by radioisotopes. *Proc. Am. Hort. Soc.* **71**: 496–506.

Tyler, G. (1971). Studies on the ecology of Baltic sea shore meadows IV. Distribution and turnover of organic matter and minerals in a shore meadow ecosystem. *Oikos*. **22**: 265–91.

Tyler, G. (1972). Heavy metals pollute nature, may reduce productivity. *Ambio*. **1**: 52–9.

Tyler, G. (1978). Leaching rates of heavy metal ions in forest soil. *Water Air Soil Pollut*. **9**: 137–48.

Tyler, G., C. Gullstrand, K-Å. Holmquist and A.-M. Kjellstrand (1973). Primary production and distribution of organic matter and metal elements in two heath ecosystems. *J. Ecol.* **61**: 251–68.

van de Geijn, S. C. and C. M. Petit (1978). *In vivo* measurement of cadmium (^{115}Cd) transport and accumulation in stems of intact tomato plants (*Lycopersicon esculentum* Mill.). II. Lateral migration from the xylem and re-distribution in the stem. *Planta*. **138**: 145–51.

van Hook, R. I., W. F. Harris and G. S. Henderson (1977). Cadmium, lead and zinc distributions and cycling in a mixed deciduous forest. *Ambio*. **6**: 281–6.

Vostal, J. (1972). Transport and transformation of mercury in nature and possible routes of exposure. In: *Mercury in the Environment* (L. Friberg and J. Vostal (eds)), CRC Press, Cleveland, pp. 15–27.

Wallace, A., E. M. Romney and J. Kinnear (1977). Frequency distribution of several trace metals in 72 corn plants grown together in contaminated soil in a glasshouse. *Commun. Soil Sci. Pl. Anal.* **8**: 693–7.

Wedding, J. B., R. W. Carlson, J. J. Stukel and F. A. Bazzazz (1975). Aerosol deposition on plant leaves. *Environ. Sci. Technol.* **9**: 151–53.

Whitby, L. M. and T. C. Hutchinson (1974). Heavy metal pollution in the Sudbury mining and smelting region of Canada. II. Toxicity tests. *Environ. Conserv.* **1**: 191–200.

Whittaker, R. H., G. E. Likens, F. H. Bormann, J. S. Eaton and T. G. Siccama (1979). The Hubbard Brook ecosystem study: Forest nutrient cycling and element behaviour. *Ecology*. **60**: 203–20.

Wieser, W., G. Busch and L. Büchel (1976). Isopods as indicators of the copper content of soil and litter. *Oecologia, Berl.* **23**: 107–14.

Witkamp, M. and B. S. Ausmus (1976). Processes in decomposition and nutrient transfer in forest systems. In: *The Role of Terrestrial and Aquatic Organisms in Decomposition Processes* (J. M. Anderson and A. Macfadyen (eds)), Blackwell, Oxford, pp. 375–96.

Woodwell, G. M., J. T. Ballard and E. V. Pecan (1975). Ecological succession and ionic leakage in terrestrial ecosystems. In: *Symposium Proceedings, Vol. II, International Conference on Heavy Metals in the Environment* (T. C. Hutchinson (ed)), Toronto, Canada, pp. 189–98.

CHAPTER 4

Impact of Metals on Ecosystem Function and Productivity

M. H. MARTIN and P. J. COUGHTREY
Department of Botany, University of Bristol, UK

1. INTRODUCTION

An ecosystem is the assemblage of biotic components, plants, animals and microorganisms, with the abiotic, physico-chemical environment to form a self-contained entity. In his definition, Tansley (1935) emphasized the importance of interactions between the components of the ecosystem; a concept which led to a new impetus in ecological thought and research in attempting to understand the structure and dynamics of such systems.

The realization that living systems are in a stable state of dynamic equilibrium, resulting from the interaction of numerous factors and processes, has given much greater insight into the effects of environmental stresses on ecosystems. Pollution of an ecosystem may be regarded as a stress imposed on the system. The effect of such a stress depends on a number of features including; type of pollutant, degree of stress imposed (intensity and duration) and the resilience or response of the most sensitive components.

The relationship between ecosystem diversity and stability has received much attention and the oft-quoted dictum that high species diversity imparts greater resilience (homeostasis) against change or stress has been largely rejected, on the basis of evidence from theoretical modelling of ecosystems, and from practical studies. Indeed, it is now suggested (May, 1976, 1978) that complex ecosystems are dynamically fragile; being less resistant to disturbances imposed on them by man than relatively simple dynamically stable systems. Odum (1975) points out that high diversity systems under stress result in lower diversity, while systems of low diversity may actually increase in diversity under the influence of pollution or other perturbations.

The general effects of pollutants on ecosystems have been considered by Bordeau and Treshow (1978), Craig and Rudd (1974), Regier and Cowell (1972), Stickel (1975) and Woodwell (1970), all of whom have taken the view that the effect of a pollution stress on an ecosystem will progressively reduce species diversity and subsequently have adverse effects on productivity. Such changes in species diversity will be manifest by a reduction in species of stable habitats together with a tendency towards *K*-selection strategies, but may allow an increase in the proportion of opportunist species with *r*-selection tendencies (Southwood, 1976). Holdgate (1979), discussing the effects of pollutants on ecosystems in general terms, has suggested a sequence of events which are associated with increasing pollutant impact. He refers to pollutants having a cascade effect in which the following stages may be recognizable:

(1) Individual pollutants may cause some detectable biochemical changes which the organism may counteract by detoxification, biochemical repair or excretion.
(2) Physiological effects which may cause growth rates to decline and affect reproduction.
(3) Impairment of competitive vigour leading to effects on populations such that the least resilient species decline.
(4) Ecological effects caused by alteration in the composition and balance of communities which eventually may lead to ecosystem disruption.

Clearly the impact on ecosystems of pollutants in general, and of heavy metals in particular, cannot be specified in detail, because each ecosystem type may react differently. The effects of individual pollutants may also result in different responses. The complexity of ecosystems and ecosystem responses are such that both our understanding and detailed knowledge of individual ecosystems are relatively incomplete. Thus, in the majority of instances, pollutant impact on ecosystems is difficult to define and is usually couched in generalities.

2. OCCURRENCE OF HEAVY METALS IN TERRESTRIAL ECOSYSTEMS

Freedman and Hutchinson (Chapter 2) have considered the sources of heavy metals in the environment while Volume 1 has examined the occurrence, concentrations and effects of individual heavy metals in plants. In an ecosystem, the occurrence and concentrations of heavy metals in any component depend largely on the source of the heavy metal. In this context the sources of heavy metals to the ecosystem will be either via the soil or by deposition of aerial particulates. In an ecosystem which has developed on a substrate rich in heavy metals, either as mining/smelting waste or naturally occurring in soil parent material, there is likely to be only one primary route of heavy metals to the ecosystem components, namely the soil. In aerially contaminated ecosystems the primary source is from aerial deposition of metal particulates, but, as the contamination continues, accumulation of metals in the soil occurs and plant uptake from the soil then becomes a further complication in interpreting the concentrations of metals found in various components of the ecosystem.

Soil parameters (e.g. pH, base status, organic matter content, cation-exchange capacity, etc.), are all known to affect the availability of heavy metals for uptake by plants (Andersson, 1976, 1977; Allaway, 1968; Bolton, 1975; Haghiri, 1974; Lagerwerff, 1967; Mahler *et al.*, 1978; Pinkerton and Simpson, 1977; Sorteberg, 1974). Furthermore, different plant species take up metals to varying degrees (Cha and Kim, 1975; John and van Laerhoven, 1972; Jarvis *et al.*, 1976; Martin *et al.*, 1980; Page *et al.*, 1972; Pettersson, 1976, 1977) and some species accumulate individual elements to very high concentrations in different organs or tissues (Antonovics *et al.*, 1971; Brooks, 1972; Ernst, 1974, 1976; Peterson, 1971; Rascio, 1977).

In ecosystems subjected to aerial deposition of particulates containing heavy metals, the concentrations of these elements in various ecosystem components are affected by both chemical form of the metal and particle size. The interaction between particle size and characteristics of different plant surfaces greatly affects the impaction and retention of the particulates (Little, 1977; Little and Wiffen, 1977; Chamberlain, 1975). During wet weather, metals may be removed by wash-off. In addition, foliar uptake may take place but will depend largely on the solubility, and hence the chemical form, of the particulates (Little, 1973; Arvik and Zimdahl, 1974; Florkowski *et al.*, 1979).

Other complications involve the distribution of metal in the soil profile; in aerially polluted ecosystems the superficial layers of soil contain high concentrations of metals while at greater depth the concentrations progressively approach background levels for the site in question. In such situations shallow-rooting species are liable to take up metals from the soil to a greater extent than deeper rooted species. In contrast, in ecosystems with high concentrations of metals in the soil derived from underlying mineralized material, parent material concentrations usually increase with

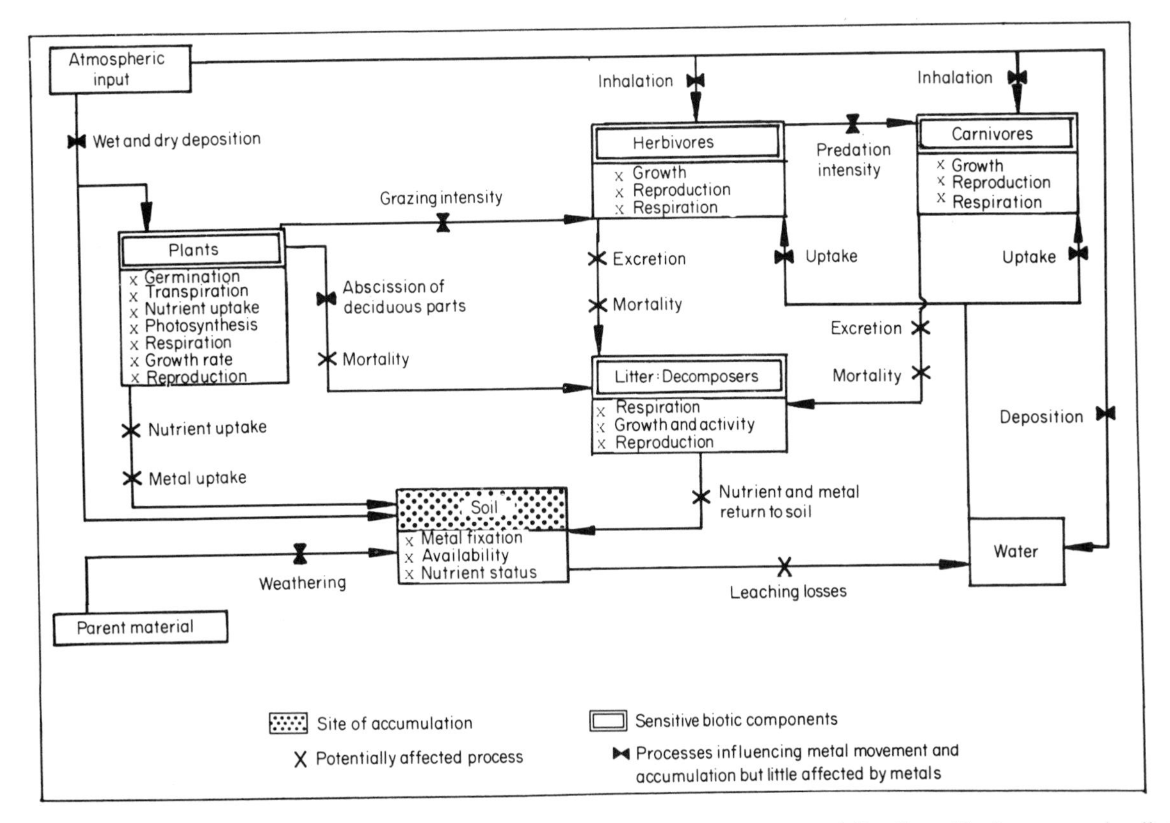

FIG. 4.1 Pathways of heavy metal uptake in ecosystems and processes which are potentially affected by heavy metal pollution.

increasing soil depths and deeper rooted species may take up more or less metal than shallow-rooted species, depending on the availability of the metal.

The consequences of heavy metals in terms of terrestrial ecosystem function are summarized in Fig. 4.1. This shows the generalized trophic levels which occur in an ecosystem and the various physiological and biological processes which may be affected by heavy metals, in terms of reducing physiological processes, which in turn can affect productivity, fecundity and mortality. At the same time Fig. 4.1 indicates the potential pathways of accumulation of heavy metals in a terrestrial ecosystem.

Figure 4.2 shows an example of the authors' work demonstrating the concentrations of Cd, Pb, Zn and Cu in a woodland ecosystem situated 3 km downwind of a primary lead/zinc smelter. These data emphasize that even in an aerially polluted system, the major sink for all heavy metal emissions is the soil. Thus, any effects on ecosystem structure and productivity are likely to be manifest, in the first instance, on the soil and litter microcosm (Tyler, 1972).

Such a conclusion is supported by the data of Smith (1976) for roadside ecosystems and by Andren *et al.* (1973), Denaeyer-De Smet (1974), Denaeyer-De Smet and Duvigneaud (1974), Heinrichs and Mayer (1977), Parker *et al.* (1978), Yost (1978, 1979), van Hook *et al.* (1980) and Tyler (1970) for terrestrial ecosystems variously polluted with aerial fallout of lead, zinc, cadmium and copper. The transfer of metal from one ecosystem component to another has only been rarely studied. The major part of this research can be ascribed to van Hook and co-workers (Blaylock *et al.*, 1973; Huckabee and Blaylock, 1973; van Hook *et al.*, 1976, 1977, 1978) who initially studied transfer rates of mercury and cadmium using radioactive isotopes in experimental situations. Later, van Hook *et al.* (1977, 1978, 1980) provided information on the transfer of cadmium, lead, zinc and copper in a deciduous mixed forest. Comparable data for the site represented by Fig. 4.2 are shown in Fig. 4.3. It is apparent from these data and those referred to above that the soil represents the major depository of the total ecosystem metal pool and that current annual transferences to this component are small, in comparison to that proportion already present. Such data have yet to be considered from the mathematical point of view to determine the long-term transfer of ecosystem metal loads; this represents an area where further research is required before ecosystem metal accumulation and distribution can be considered in any other context than the present. With the advent of more precise field data (as demonstrated in

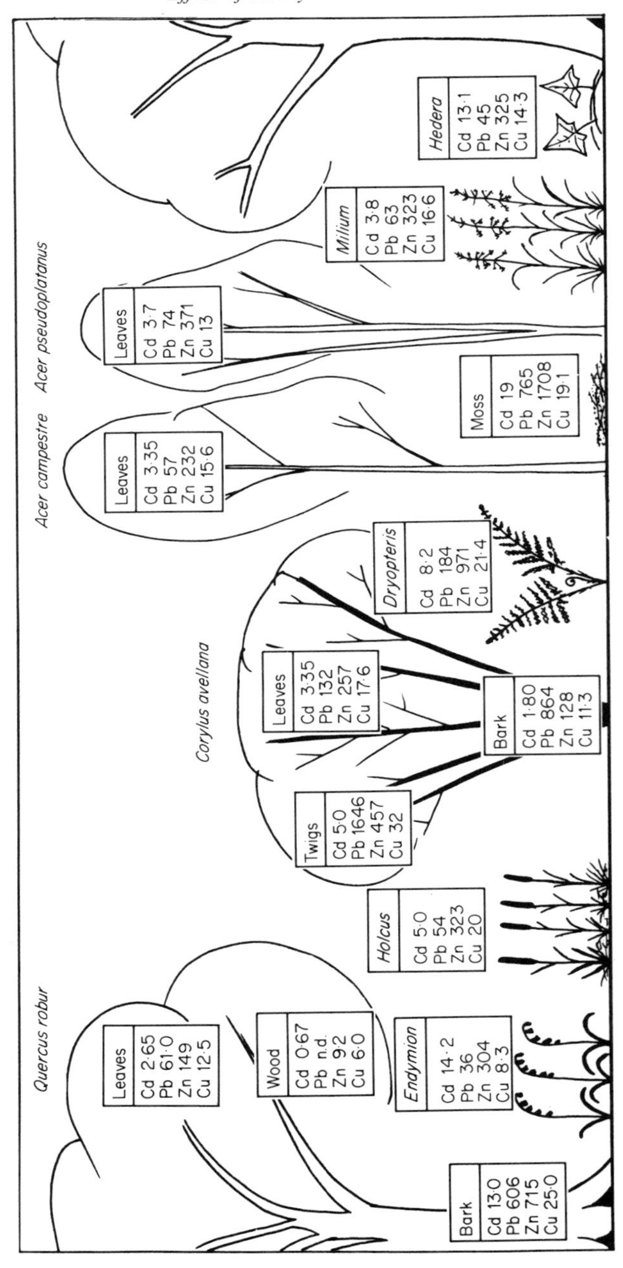

FIG. 4.2 Concentrations and distribution of heavy metals in a deciduous woodland ecosystem situated 3 km downwind of a primary lead–zinc smelter at Avonmouth, England. All concentrations in μg/g (dw), soil levels for concentrated nitric acid extracts.

	Cadmium	*Lead*	*Zinc*	*Copper*	*pH*	*% C*
Litter	*44·46*	*2 559*	*2 267*	*133*	*3·80*	*64·2*
Soil (depth, cm)						
0–1	22·89	1 432	1 010	56·4	4·00	12·6
1–3·5	16·21	370	640	23·4	4·10	5·63
3·5–6	18·95	88	689	17·4	4·15	3·87
6–8·5	18·03	66	846	16·6	4·50	2·70
8·5–11	14·20	47	673	15·9	4·70	2·01
11–13·5	12·87	50	619	15·8	4·85	2·09
13·5–16	7·70	42	485	16·6	4·70	2·12
16–18·5	6·77	41	513	17·1	4·65	2·04
18·5–21	5·30	33	489	16·9	4·50	1·51
21–31	3·77	35	440	23·4	4·75	1·69
31	2·24	15	401	24·9	5·15	1·30

Species	*Soil and litter dwelling invertebrates* *Cadmium*	*Lead*	*Zinc*	*Copper*
Clausilia bidentata	53·0	245	613	—
Helix aspersa	52·5	39·0	404	87
Oniscus asellus	232	568	462	459
Arion hortensis	57·0	63·6	1 430	—
Arion fasciatus	30·0	118	1 250	—
Lumbricus spp.	41·0	32·8	1 502	7·3

Species	*Fungi (fruiting bodies)* *Cadmium*	*Lead*	*Zinc*	*Copper*
Fistulina hepatica	4·0	17·2	93·5	37·8
Auricula auricularia	3·8	32·8	138·1	6·56
Stereum hirsutum	3·6	94·3	171·7	15·9
Daldinia concentrica	0·8	10·8	45·3	6·2

FIG. 4.2—*contd.*

Figs. 4.2 and 4.3) it should become possible to produce predictive mathematical models, possibly following the lines of that described by Garten *et al.* (1978) for the dynamics of plutonium in a contaminated deciduous forest ecosystem.

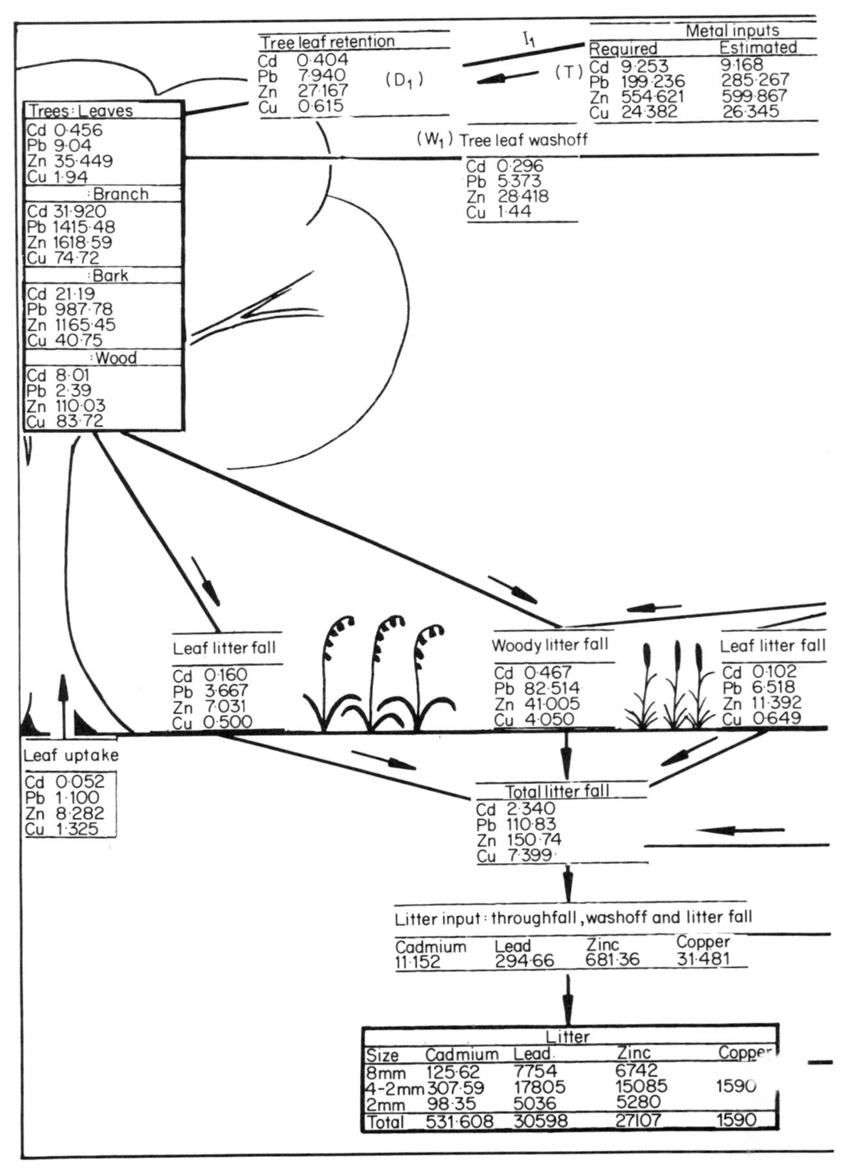

Size	Cadmium	Lead	Zinc	Copper
8mm	125·62	7754	6742	
4-2mm	307·59	17805	15085	1590
2mm	98·35	5036	5280	
Total	531·608	30598	27107	1590

FIG. 4.3 Heavy metal inputs, transfers and accumulations in a deciduous woodland ecosystem (see Fig. 4.2). Metal data are mg metal/m^2 for accumulations (□) and mg metal/m^2/yr for inputs and transfers (arrows represent direction of transfers).

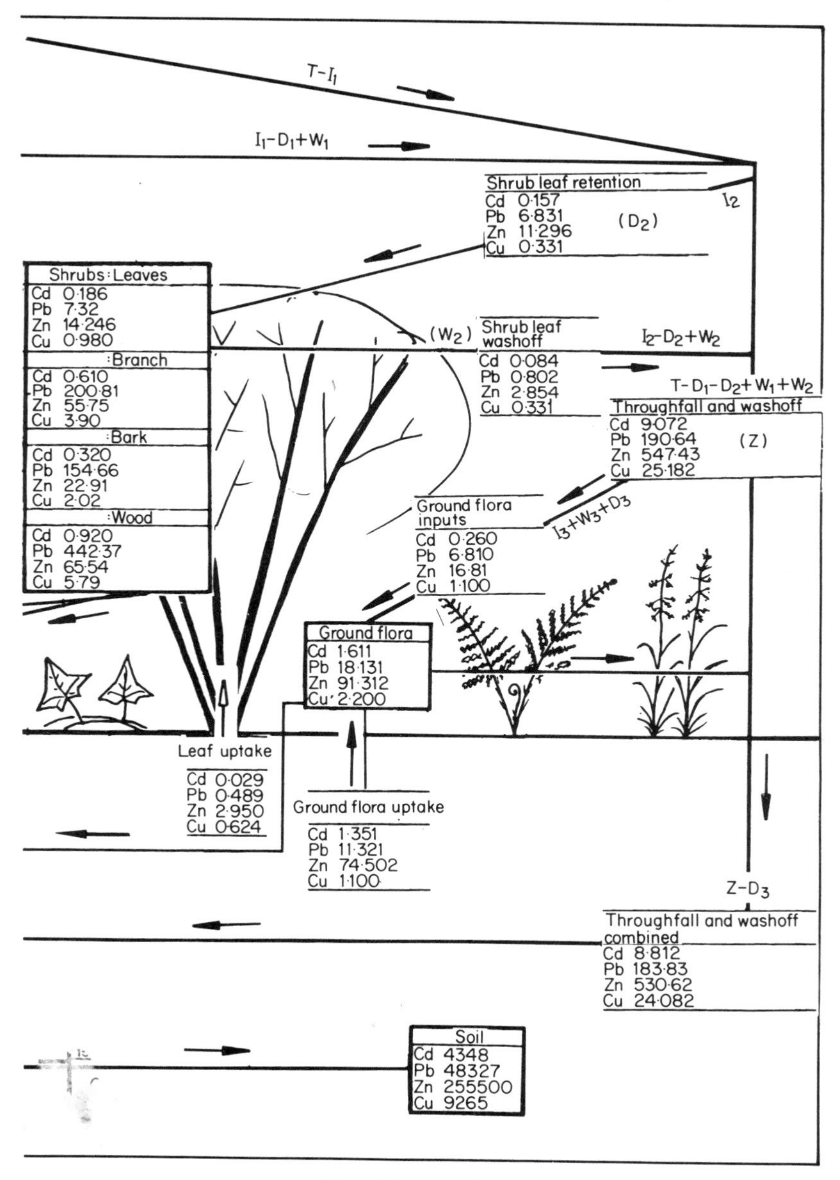

FIG. 4.3—*contd.*
T = total input to system, I = input from aerial sources to vegetation, D = retention by vegetation, W = washoff from vegetation, Z = total throughfall, washoff and dry deposition at 0·3 m above ground level.

3. FOOD CHAIN RELATIONSHIPS

The accumulation of heavy metals through the soil–plant–herbivore–carnivore food chain, allowing an intensification of their concentrations and effects, is well documented. Particular concern has arisen from the effects of mercury and thallium, both of which have been distributed in the environment via grain coated with compounds of these metals to prevent loss after sowing from small mammal predation. Mendelssohn (1962, 1972) reported mass destruction of bird life in Israel from the direct effects of contaminated grain and also by secondary poisoning of carnivorous animals. Similarly Borg *et al.* (1969) reported primary and secondary effects of alkyl-mercury poisoning in seed-eating and predatory birds in Sweden. Although mercury levels were negligible in grazing animals (e.g. hares and deer) mercury poisoning was demonstrated in mammalian predators such as foxes, martens and polecats. Generally, elevated levels of metals are found in animals in areas of high wheat production (Curnow *et al.*, 1977) and in urbanized areas (White *et al.*, 1977). Chaney (1973) has reviewed the foodchain effects of metals applied to land via sewage sludges and effluents and has suggested that the elements cadmium, copper and zinc are a potential hazard to food chains via plant accumulation.

It was apparent from earlier discussion (Section 2) that one of the areas of metal accumulation in terrestrial ecosystems is the soil–litter system. In studies of plutonium dynamics in a deciduous forest, Garten *et al.* (1978) found that less than 0·25 % of the total plutonium was residing in the biota, the soil being the main repository. It is apparent that an early food-chain effect might be expected to be associated with organisms inhabiting the soil–litter component of the ecosystem. Data concerning the accumulation of metals in soil-inhabiting organisms are fragmentary, but provide some evidence that invertebrates do accumulate metals to levels over and above that of their food supply, representing a potential hazard to carnivores. Furthermore, some metals appear to be selectively accumulated at the decomposer stage. In the case of plutonium (Garten and Dahlman, 1978) concentration ratios relative to soil for litter, invertebrate cryptozoans, herbaceous ground vegetation, orthoptera and small mammals were approximately 10^{-1}, 10^{-2}, 10^{-3}, 10^{-3} and 10^{-4}, respectively. Williamson and Evans (1972), in a study of lead levels in roadside invertebrates and small mammals, showed that lead levels in tissues of small mammals were low in comparison to the levels in many of their prey. Joose and Buker (1979) suggested that 15 % of lead uptake by *Collembola* could be lost in their exuviae and Zhulidov and Emets (1979) have shown that lead levels in beetles, although related to environmental levels, were in most species no higher than those of soil and vegetation. However, Gish and Christiensen (1973) published concentrations of cadmium, nickel, lead and zinc concentrations in earthworms from roadsides and suggested that animals fed on earthworms for extended periods could accumulate lead and zinc to toxic levels. Furthermore, Giles *et al.* (1973) suggested that lead was concentrated by insect predators during the late part of the season, and the herbivore–carnivore transfer of lead is also supported by Price *et al.* (1974).

In metalliferous and mining areas, copper content of isopods was shown to be closely related to the copper concentration of the litter on which they fed (Wieser *et al.*, 1977). Although the isopod copper content was also related to temperature, it was always greater than that of the litter. In a base-metal mining area of Wales, Ireland (1975) found that earthworms retained and accumulated lead whilst regulating iron and zinc contents. Lead accumulated by earthworms could be transferred to toads (*Xenopus laevis*) in experimental studies (Ireland, 1977). Although strip mining for coal had a minimal effect on levels of cadmium, lead and mercury in resident wild life (Curnow *et al.*, 1977), in abandoned metalliferous mines, small mammal lead concentrations were elevated relative to control areas (Roberts *et al.*, 1978). Near smelters and industrial sites the metal concentrations in invertebrates and their predators are also elevated. Thus Bull *et al.* (1977) reported high mercury concentrations in wood mice and bank voles near an industrial source, while Martin and Coughtrey (1975) reported elevated cadmium concentrations in herbivorous and carnivorous invertebrates near a smelter. The degree of cadmium elevation was related to distance from the smelter (Coughtrey *et al.*, 1977). Munshower (1972, 1977) reported cadmium concentrations in grasshoppers which were generally 2–3 times higher than in herbage close to a smelter. He also suggested that the maximum elevation of cadmium occurred at the herbivore stage of the food chain; amongst the carnivores he studied, only badgers had elevated cadmium concentrations; which could be a facet of the badger's diet.

Although there is some evidence for biological accumulation of metals through the plant–herbivore and litter–detritivore stages, it is also apparent that the degree to which this occurs is dependent on the metal concerned. Thus, although field data would support accumulation of cadmium as discussed above, in an experimental situation van Hook and Yates (1973) showed no evidence for cadmium accumulation in a vegetation–cricket–spider food chain. van Hook (1974) studied cadmium, lead and zinc concentrations in earthworms and the soils which they inhabited. By virtue of a concentration ratio defined as:

$$\frac{\text{Concentration of metal in earthworm}}{\text{Concentration of metal in soil (top 10 cm)}}$$

he suggested that there was a biological concentration of cadmium and zinc but not of lead in the species studied. Similar results were obtained by Martin and Coughtrey (1976) for a wider range of terrestrial litter-decomposing organisms (Table 4.1). This relative bioaccumulation of cadmium in comparison to lead or zinc was confirmed by reworking the data of Gish and Christiensen (1973) for roadside sites and the data of Ireland (1975) for a mine site. A similar conclusion concerning the relative

TABLE 4.1

CONCENTRATION FACTORS FOR INVERTEBRATES (AND CONCENTRATION RANGES FOR PRODUCERS/HERBIVORES) CLOSE TO A HEAVY METAL SMELTER. (ADAPTED FROM MARTIN AND COUGHTREY, 1976)

Animal	Lead	Zinc	Cadmium
Earthworms	0·34	0·45	7·60
Slugs	0·42	1·85	5·28
Snails	0·43	2·94	41·6
Woodlice	0·42	0·30	5·90
Summary of ranges of concentrations of metals in producers and herbivores (μg/g (dw))			
	Lead	Zinc	Cadmium
Producers	122–685	260–2 395	2–25
Herbivores	66–551	382–1 299	29–377
Herbivores/Producers	0·54–0·80	1·5–0·54	14·5–15·1

mobility of cadmium was reached at a later date by Roberts and Johnson (1978) and is also supported by the data of Williamson (1980) for snails from a roadside site. The mobility of cadmium may be related to both the availability of the metal in the soil and the selective accumulation within different organs of the target organism (Martin *et al.*, 1976; Coughtrey and Martin, 1976, 1977*a*; Johnson and Roberts, 1978; Roberts and Johnson, 1978). Thus, at least in the case of cadmium, the plant–herbivore and litter–detritivore stages could be of great importance in subsequent transference of the metal through the food chain. Data for other metals are rather scarce but some may be obtained by reworking the data of Sharma and Shupe

(1977*a*,*b*). Concentration factors calculated from their data for *Spermophilous variegatus* (Rock squirrel) and for soils give the following results:

Metal	*Mean c.f.* ± *S.E.*	% *Variation (S.E. of mean)*
Pb	0·25 ± 0·099	39·6
Cd	3·44 ± 0·324	9·42
As	1·74 ± 0·700	40·2
Cu	0·402 ± 0·056	13·9
Mo	1·212 ± 0·304	25·1
Se	14·708 ± 1·367	9·29
Zn	4·629 ± 0·650	14·04

These data suggest a greater and more consistent bioaccumulation of selenium, zinc and cadmium relative to lead and copper, with lead, arsenic and molybdenum showing much greater variation.

In comparison to data for metal accumulation through the lower steps of the terrestrial food chain, data for actual effects are much more restricted. Data for litter-decomposing invertebrates are discussed elsewhere (Section 4.2.1) but in the present context, experimental studies with mercury (Haney and Lipsey, 1973) suggested that aphids feeding on mercury-contaminated tomatoes showed a significant decrease in fecundity. van Rhee (1975) demonstrated the toxic effects of copper on earthworms in pastures contaminated with pig wastes. In experimental situations this author also found that copper and mercury reduced body weight of earthworms, caused the loss of the clitellum and reduced cocoon production. Thus the effects of metal bioaccumulation through food chains may not be immediate in causing the death of an individual organism but may reduce the breeding ability of a population.

One of the best documented effects of food chain accumulation of heavy metals concerns the soil–plant–herbivore system in agricultural practices. The toxicity of mine wastes to grazing animals was reported as early as 1919 (Griffiths, 1919). Sheep, horses and poultry were most susceptible while effects on cattle were less pronounced. However, Allcroft (1951) showed that chronic lead poisoning was unlikely to occur in animals grazing pastures contaminated by galena or plumbiferous dusts and that the hazard of acute poisoning from leaded paint was much greater. Although this may be true for lead in cattle, reports for other animals and other metals suggest that contamination of pastures can cause toxic effects. Gracey and Todd (1960) reported that six out of 95 ewes died on a pasture that had been sprayed with a copper sulphate molluscicide for liver fluke control.

Stokinger (1963) reported poisoning of cattle and horses within five miles of a zinc and lead smelter. Donovan *et al.* (1969) reported high levels of lead in the kidneys of animals that had died while grazing a pasture adjacent to a mine in Ireland. Maga and Hodges (1972) reported deaths of horses in particular and also cattle and sheep close to a smelter. Factors involved in soil–plant–animal relationships in agricultural situations were discussed by Alderman (1968). Five factors were highlighted:

(1) The amount of dry matter intake (variable according to pasture).
(2) Species composition of forage.
(3) Fertilizer regime.
(4) Soil type and pH.
(5) Intake of trace elements from direct animal foods.

Two other factors should also be considered, that of direct soil ingestion, and that of trace element interaction. Considering the former, Healey (1968, 1974) suggested that cattle could ingest 91–454 kg of soil/annum/cow with peaks in autumn and winter, while sheep ingested 2 kg of soil/week. Moreover, herbage from grazed areas could contain up to 25 % (dw) of soil which could be a source of cobalt, manganese, selenium and zinc. Hence a direct soil–animal stage should also be considered in these agricultural situations.

Toxic effects resulting from food chain accumulation are complicated by diverse interactions. An example of this complexity is that of molybdenum–copper–zinc–lead interactions in cattle. According to Dick and Bull (1945) a small amount of molybdenum may bring about a copper deficiency in grazing animals Thus, in a study of 'S-Charl disease' of cows and calves, high levels of molybdenum were found to be the most important factor (Hogl, 1975). This was similar to studies of teart pastures in Somerset, England, where Ferguson *et al.* (1940) found the cause of a comparable disease to be high levels of molybdenum, the effects of which could be rectified by either feeding the animals, or spraying the pastures, with copper sulphate. Similarly Vetter and Mahlop (1971) suggested that zinc harmed animals via a reduction in copper absorption, but the effect of zinc contamination near a smelter was not so severe as expected because copper was also a component of the emissions. In horses, Roberts *et al.* (1974) suggested that cadmium reduced zinc absorption and Willoughby *et al.* (1972) suggested that toxic amounts of zinc could prevent the development of clinical signs of lead poisoning. It is thus apparent that the effects of food chain accumulation of metals in sites contaminated with several elements could be less pronounced in their initial stages unless the normal ratios and balances of different metals have become disturbed. Thus, Griffiths and Wadsworth (1980) state in context of lack of thrift in young stock close to a smelter at Avonmouth, England, that 'the interaction of elements, particularly on the copper status of the animals is indicated in recent research findings, and in the area of this investigation the combined effects of lead, zinc, cadmium and molybdenum can be postulated'.

In conclusion, metal contamination can, via food chain accumulation, cause excessive levels of certain metals in some animals and hence toxic effects. Some metals, e.g. cadmium, appear to be relatively mobile within terrestrial food chains but subsequent mobility of any one metal, once ingested by a herbivorous animal, depends on the site of accumulation within the body tissues. Although food chain accumulation of metals may not itself cause the death or injury of an individual organism it can cause a reduction in the breeding potential of a population. Metal contamination in agricultural ecosystems can cause a reduction in secondary productivity but conclusions are complicated by the extremely complex nature of interactions between metals within the target organism; indeed, Dorn *et al.* (1975) state 'It would be prudent to consider the possibility of livestock and food chain contamination in the location, design and operation of smelters'. The final step in all agricultural food chains is man; as a result, the movement of potentially toxic heavy metals through agricultural food chains should be viewed with concern.

4. ACTUAL CASE STUDIES

4.1. Field Evidence for Ecosystem Effects

Tyler (1970), in a study of lead in a roadside ecosystem, showed that 95 % of the non-lattice lead was located in the mor layer and the mineral soil. Accumulation of metals in the soil plus litter layer is one of the best-documented ecosystem effects of heavy metals, and Tyler (1972) suggests that 'possible effects of heavy metals in biological systems will first occur at the destruent level, which contains the organisms responsible for the decomposition of litter, and symptoms will appear in a slower or less complete humification and remobilization of the minerals'. He thus reports (Tyler, 1975*a*) that copper and zinc pollution around a brassworks brought about an accumulation of incompletely decomposed litter, particularly evident where the conifers were damaged or dead. Similarly, Jordan and Lechavalier (1975) and Strojan (1978*a*) reported accumulations of litter in O_2 horizons close to a metal smelter at Palmerton, Pennsylvania. Watson

(1975) also reported accumulations of litter up to 7·4 km from the stack of a primary smelter in south-east Missouri, and Coughtrey *et al.* (1979) discussed similar results around a metal smelter at Avonmouth, England. In the latter case, multiple regression analysis supported the hypothesis that litter accumulation was closely related to concentrations of cadmium in particular, and also to lead, copper and zinc. Furthermore, litter accumulation was associated with a particular size range of materials not including freshly fallen leaves.

The hypothesis that reduced decomposition occurs close to heavy metal sources is supported by studies of the rate of decomposition of litter enclosed by mesh bags (Inman and Parker, 1978; Strojan, 1978*a*). Percentage decomposition of bagged litter in two woodlands close to the smelter at Avonmouth were $7{\cdot}64 \pm 2{\cdot}87$ and $2{\cdot}40 \pm 1{\cdot}10$ compared to $31{\cdot}44 \pm 4{\cdot}17$ and $14{\cdot}12 \pm 0{\cdot}42$ in uncontaminated woodlands during a nine-month period from December to September (Coughtrey *et al.*, unpublished). Evidence for decreased decomposition comes also from the work of Cole and Turgeon (1978) in sites affected by calcium arsenate sprays and from the work of Williams *et al.* (1977) in a site affected by past mining activities. However, where the scale of contamination is much less, e.g. urban areas, Parker *et al.* (1978) found that the total dry weight of litter/unit area was less than had previously been published for similar vegetation types. Similarly, Lawrey (1977, 1978) found no evidence for decreased decomposition in sites previously involved in strip-mining for coal.

Field evidence for reduced litter decomposition in heavy metal contaminated areas is recent in origin, however. As early as 1914 Lipman and Burgess showed that copper, lead, zinc and iron could reduce ammonification and stimulate nitrification in soils. Furthermore Nielson (1951) and van Rhee (1975) reported adverse effects of soil minerals and copper on earthworm populations. Field studies of microorganisms and invertebrates are also limited; Jordan and Lechavalier (1975) reported that total numbers of bacteria, actinomycetes and fungi were greatly reduced in soils heavily contaminated with zinc; and Strojan (1978*b*) reported that densities of some invertebrate groups were lower close to the same smelter. A particular effect was noted on the Oribatid mites, and a similar effect was noted by Williams *et al.* (1977) in an ex-mining site and by Fairley (unpublished) in a smelting site. At the time of writing it is not clear whether the increased amounts of litter in metal contaminated sites are related either to an effect on microorganism populations or invertebrate populations; available data are often fragmentary and conflicting. Problems arise in expression of results with studies of invertebrate numbers and density for sites where the total amount and density of the litter is much greater. Furthermore, Jordan and Lechavalier (1975) conclude that the reduction in microbial populations near the Palmerton smelter may be a cause of reduced litter decomposition while Strojan (1978*b*) implies that the reduction in invertebrate numbers is the cause.

At Avonmouth, England there is no evidence for a reduced soil microbial population at sites where litter accumulation occurs (Shales, unpublished) although the same litter concentrations can have an adverse effect on litter organisms, e.g. isopods (Coughtrey, unpublished). However, Coughtrey *et al.* (1980*a*) have tentatively suggested that an action of heavy metals may take place at the microorganism/invertebrate interaction stage. Furthermore, at Palmerton, the situation concerning litter decomposition is complicated by fire (Jordan, 1975), and at many sites contaminated by heavy metals from industrial processes there are other contaminants present. Of these, the oxides of sulphur and nitrogen can create conditions of high acidity (i.e. Hutchinson and Whitby, 1977) although there is little evidence for a close correlation between low pH and increased litter accumulation in metal contaminated sites (Coughtrey *et al.*, 1979). Kelly and Henderson (1978) studied the effects of additions of superphosphate and urea to plots within a deciduous forest and found that increased nitrogen (from urea additions) decreased soil invertebrate populations by 30% while increasing decomposition rates, in contrast to phosphorus which decreased bacterial populations but also decreased decomposition rates. These authors conclude that soil acidity changes appeared to be primarily responsible for the observed changes in decomposition rates. Hutchinson and Whitby (1977) suggested that a decrease in pH increased mobility and solubility of heavy metals and thus created phytotoxic problems, hence any field study of the effect of heavy metals on soil processes should (ideally) take into account all the factors concerned. Furthermore, Tyler (1978) showed that the degree of metal availability in soil was related to the amount and acidity of rainfall, while Rutherford and Bray (1979) showed a correlation of drainage status with elevated amounts of available nickel, copper and sulphate. Finally, De Leval and De Monty (1972) suggested that the microflora of zinc and lead contaminated soils around Liège, Belgium, was dependent upon the vegetation type present rather than soil metal concentrations, although changes in vegetation were related to increased metal concentration in soil.

4.2. Experimental Evidence for Ecosystem Effects

Broadly speaking, this evidence may be separated into two types: (a) studies of experimental materials transferred from the field to the laboratory with as little disturbance as possible and with no artificial additions of heavy metals, and (b) those studies concentrating on individual ecosystem processes involving artificial additions of metals to gain a measurable response.

4.2.1. Studies on naturally contaminated materials

(*1*) *Litter decomposition and nutrient cycling*. Tyler (1975*a*,*c*) studied the effects of copper and zinc contamination on mor samples from Ostergotland by incubation of field samples in laboratory conditions. He showed that there was a strong negative correlation between total carbon loss (as carbon dioxide) and combined copper plus zinc concentrations of the samples. Although the ammonification rate appeared to be reduced close to the brassworks, no effect was noted on nitrification rates. At an equal degree of water saturation (Tyler, 1975*b*), heavy metals mainly limited the decomposition rate (as measured by loss of weight of incubated samples). A water content corresponding to less than 20% of the field capacity was required to bring about a greater reduction in decomposition rate than was caused by heavy metals. Increasing metal content decreased extractable phosphate and Tyler (1976) concluded that effects on decomposition rate and P mineralization were more closely correlated to copper and zinc concentrations, although at higher pH the heavy metal effect may have been counteracted.

Microcosms were evaluated in terms of their effectiveness for assessing pollution problems by Draggan (1976) and have been used recently to assess the effects of heavy metals on ecosystems. Lu *et al.* (1975) applied sewage sludge to model ecosystems and suggested that lead, zinc, cadmium and copper could be mobilized into ecosystem components; these authors also noted the importance of seeking the relevance of model studies to practical field studies. Jackson *et al.* (1978*a*,*b*) applied contaminated litter to forest microcosms and demonstrated mobilization of metals in soils with a sustained increase in leaching of calcium and nitrate–nitrogen in amended microcosms. Metal addition via contaminated litter increased the daily and cumulative gaseous carbon loss (Ausmus *et al.*, 1978). Similar effects on carbon loss were demonstrated by Chaney *et al.* (1978) in microcosms contaminated with litter and soil from East Chicago.

(*2*) *Enzyme activity*. Apart from the work of Tyler (1975*b*, 1976) and Ruhling and Tyler (1973) there are few published studies of enzyme activity in natural substrates contaminated with heavy metals, studied under laboratory conditions. Tyler showed that phosphatase activity was decreased with increasing heavy metal content while β-glucosidase activity was not (except possibly in the case of lead). Urease activity was decreased at very high concentrations of metals. Varanka *et al.* (1976) reported the effect of sewage sludge additions on soil dehydrogenase, invertase, urease, cellulase, amylase and protease activities. Protease and amylase activities increased while urease and invertase activities decreased with sludge treatments. No attempt was made to correlate these effects with the increased metal contents of soils treated with sewage sludge.

3. Microorganisms and invertebrates. Jordan and Lechavalier (1975) reported that bacteria, actinomycetes and fungi, tolerant to zinc incorporated in dilution plate media, could be isolated from soils surrounding the Palmerton smelter. Similar conclusions concerning metal tolerance in microorganisms were reached by Tatsuyama *et al.* (1975*a*), Hartman (1974), Uchida *et al.* (1973), Coughtrey *et al.* (1980*b*). As a result, the concept of using microorganisms as indices of environmental heavy metal pollution was put forward by Balicka *et al.* (1977). Adaptation to metal stress in microorganisms is often via a change in biochemical factors initially adversely affected by the metal concerned, e.g. selenium (Letunova, 1970), or mercury (Tonomura *et al.*, 1968), and can be important in changes on metal speciation in soil (e.g. Kimura and Miller, 1964). Experimental data for either microorganisms or invertebrates in artificial systems contaminated by existing natural materials are sparse; although in microcosm experiments Inman and Parker (1978) showed no consistent effect in fungal populations of materials from a clean and contaminated site.

van Rhee (1975) included experimental data for decrease in earthworm fecundity as a result of copper contamination from disposal of harbour muds, and Coughtrey (unpublished) has shown that isopods fed contaminated litter suffer increased mortality as a consequence of increased metal uptake. Coughtrey *et al.* (1980*a*) have also shown that one consequence of feeding *Oniscus asellus* on litter collected from a metal-contaminated site was an increase in metal tolerance of the gut microflora with a decrease in the numbers of actinomycetes relative to bacteria.

4.2.2. Studies on artificially contaminated materials

(*1*) *Litter decomposition and nutrient cycling.* Lu *et al.* (1975) added lead chloride and cadmium chloride to soils to give final concentrations of 10 ppm Cd or Pb and then included these soils as substrates within a microcosm system containing both aquatic and terrestrial components. The cadmium treatment was shown to be more toxic in these systems than was the lead treatment. Results obtained were said to be similar to those of field experiments. A similar toxic effect was demonstrated by Bond *et al.* (1976) when additions of 10 ppm Cd to a coniferous soil microcosm inhibited oxygen uptake and reduced carbon dioxide output. More recently, Chaney *et al.* (1978) showed interactions between cadmium and zinc by additions of all combinations of 0, 0·1, or 10 ppm cadmium chloride and 0, 100 or 1000 ppm zinc chloride to microcosms containing litter and soil from an uncontaminated site. Reduced respiration rates were detected at high concentrations of either cadmium or zinc. Suppression of respiration was not related to excess Cl^- ion since additions of potassium or calcium chloride caused either no effect, or a stimulation of respiration. A comparison was also made with naturally contaminated microcosms, and the authors state—'These observations raise the larger and still unanswered question of what relationship does added soluble Cd^{2+} and Zn^{2+} have to particulate inputs and what comparisons can be made between the one time additions of metals and the chronic build up situation'. Notwithstanding these comments, other authors have continued experimentation along similar lines and studied the effects of direct additions of metal chlorides to experimental units. Spalding (1979) applied mercury, cadmium, lead, nickel, zinc and copper chlorides at rates of 0, 10, 100 and 1000 ppm and calcium chloride at 7, 68 and 683 ppm to Douglas Fir needle litter and monitored rates of carbon dioxide efflux. At the highest rate of treatment, all metals, except lead and including calcium, reduced carbon dioxide output, mercury being the most potent inhibitor. Thus, to a certain extent, experimental results confirm the effect of metals on decomposition rates (measured in terms of carbon dioxide flux). However, care is required in extrapolating experimental results to field conditions, not least because of differences in availability and form of the metals between experimental and field conditions.

Experimental studies of the additions of heavy metals to soils on nutrient cycling were discussed in 1914 (Lipman and Burgess, 1914). Additions of copper, zinc, iron and lead salts to soils in an incubation experiment showed copper to be the most toxic sulphate to ammonification processes, causing a reduction of about 30%. On the other hand, the same metals caused a stimulation of the nitrifying flora, even at a concentration of 0·15% (except in the case of lead). Premi and Cornfield (1969), in a study of the action of 100 ppm and 1000 ppm copper, zinc and chromium on mineral nitrogen levels in a sandy loam soil, showed that both levels of copper and zinc and the higher level of chromium increased nitrogen mineralization, and nitrification was reduced by the higher level of copper and both levels of zinc. Wilson (1977) studied the effects of additions of zinc to three soils at rates of 0, 10, 100 and 1000 ppm and demonstrated that total inhibition of nitrification occurred at 1000 ppm in all three soils, and at 100 ppm in two out of three soils. Wilson concluded that care should be taken not to apply zinc-containing materials indiscriminately to soils. Tyler *et al.* (1974) examined the effects of cadmium, lead and sodium salts on nitrification in a mull soil from a *Filipendula ulmaria* meadow. Cadmium (both as chloride and acetate) significantly increased nitrate accumulation and these authors suggest that this would affect the nitrogen supply to the nitrifiers. They further suggest that the general soil microflora may be less tolerant to metal ions (especially cadmium) than the nitrifying bacteria, a circumstance decreasing the immobilization of mineral nitrogen.

Another effect of heavy metals on ecosystem nitrogen fixation may occur at the root nodule. According to Hallsworth *et al.* (1960), the addition of cobalt causes root nodules to increase in size, even when plants are grown on a high nitrate supply, resulting in the production of more leaves. Such a metal-sensitive system could be expected to be affected by the presence of other more toxic metal ions. Iron supply could also be altered by the toxic effects of heavy metals on iron-oxidizing bacteria (see, for example Imai *et al.*, 1975).

(2) *Enzyme activity.* The effect of lead additions to soils on enzyme synthesis was studied by Cole (1977). At a concentration of 2000 ppm Pb, and after amendment with maltose or starch, net synthesis of amylase and α-glucosidase was reduced by 75 % and 50 %, respectively. Amylase activity was much less sensitive to lead than was amylase synthesis. The degree of reduction in enzyme synthesis was related to the chemical form of the lead added, although highly insoluble lead sulphide still provided a considerable reduction in enzyme synthesis (50 % at 1720 ppm); thus, even insoluble lead compounds could act as potent modifiers of soil biological activity. A similar overall effect via enzyme synthesis rather than activity was reported by Spalding (1979) as a result of treatment of Douglas Fir needle litter with mercury, cadmium, lead, nickel, zinc and copper chlorides; after four weeks of incubation, mercury and cadmium decreased cellulase activity, mercury depressed xylanase activity, cadmium and lead depressed amylase activity, mercury and copper depressed β-glucosidase activity and cadmium stimulated β-glucosidase activity. The decline in overall respiration noted was suggested to be the result of decreased enzyme synthesis associated with inhibited microbial respiration. Tabatabai (1977) studied the effect of trace elements on urease activity in soil and found that at comparable molarities trace elements showed the following series of increased inhibition:

$$Mn^{2+} < Sn^{2+} < Zn^{2+} < Cd^{2+} < Cu^{2+} < Hg^{2+} < Ag^{2+}$$

(3) *Microorganisms and invertebrates.* There are several publications on

the effects of heavy metals on microorganisms, both in artificially contaminated media and in soils contaminated with metal compounds. Doyle *et al.* (1975) showed that species differed in their uptake of cadmium, in a way that was related to its toxicity. Gullino and Fiusello (1976) studied the toxicity of lead nitrate to 80 strains of fungi and showed that some species were more tolerant than others, tolerance and toxicity being related to dry weight production of mycelium. According to Babich and Stotsky (1977) fungal sporulation was inhibited by cadmium prior to mycelial growth, while actinomycetes appeared, in general, to be more tolerant than eubacteria. The toxicity of cadmium to eubacteria, actinomycetes and fungi was shown to be dependent on the pH of the test media. In experimental situations Gingell *et al.* (1976) showed that a combined contamination of lead, zinc and cadmium on cabbage leaves caused an effect on the inoculated phylloplane microflora. In litter, Inman and Parker (1978) reported that cadmium chloride had little effect on microfungal populations apart from the case of *Aureobasidium* sp. It is apparent from these studies and those discussed earlier (Section 4.2.1) that micro-organisms differ in their response to heavy metals both according to species and metal. Furthermore, in natural situations the soil type and pH can affect the response of microorganisms to contamination. Sewage sludge studies (Malaney *et al.*, 1959; Cenzi and Morozzi, 1977) represent an important area for effects of heavy metals on soil microorganisms. There is evidence that strains of microorganisms intimately associated with metals can show considerable tolerance to these metals. Furthermore, it is also apparent that organisms isolated from soils contaminated with one metal can show tolerance to other metals (Tatsuyama *et al.*, 1975*b*). Ecologically, such an effect may have severe consequences in that many such organisms can continue their activities in sites heavily contaminated with other metals, the presence of an active microflora then effecting the release of further soluble metal species (e.g. Ausmus *et al.*, 1977). In the long term, there is evidence that some metal-tolerant micro-organisms are also cross-tolerant to antibiotics (Timoney *et al.*, 1978; Varma *et al.*, 1976). It is necessary to repeat the proviso that it is unclear how applicable such results, obtained from application of metal salts to cultures of microorganisms, can be in interpreting possible consequences of heavy metal pollution in naturally contaminated field sites.

At the time of writing, there is very little experimental evidence for the effects of heavy metals on terrestrial soil invertebrates. Coughtrey (1978) suggested, from feeding experiments using salts of cadmium, lead and zinc, that weight gains of *Oniscus asellus* (a terrestrial isopod) were adversely

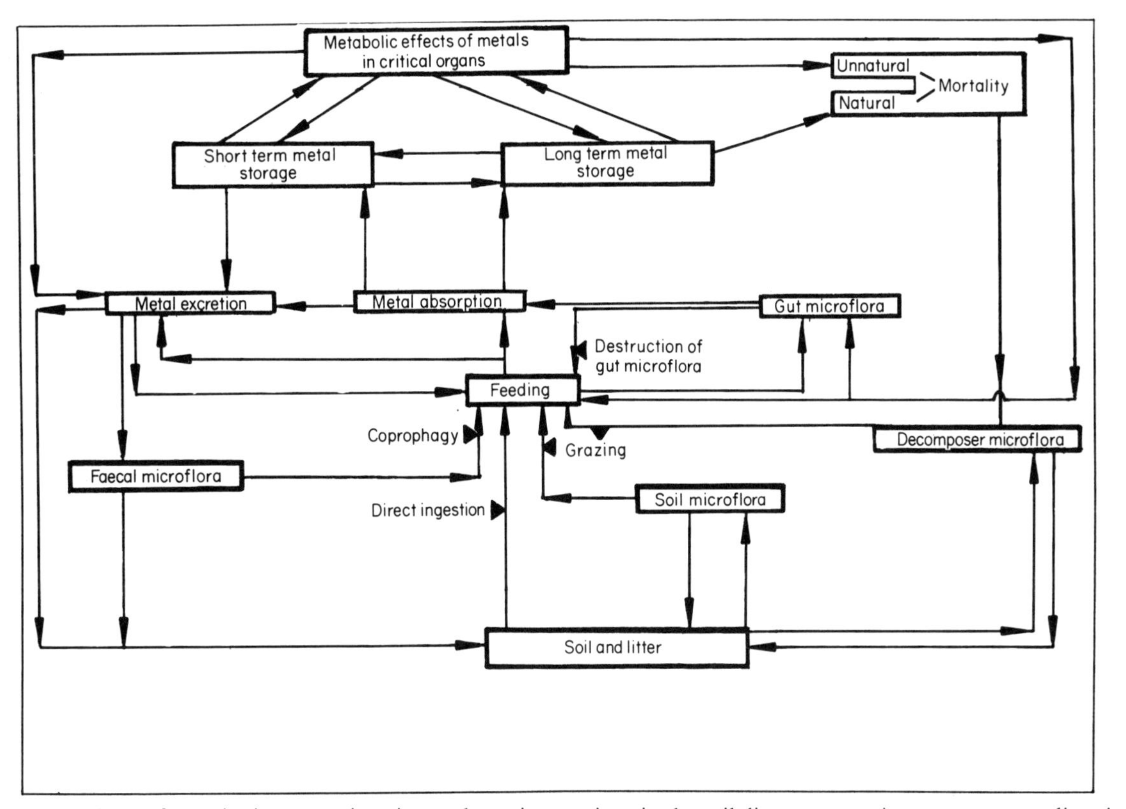

FIG. 4.4 Metal transfer and microorganism–invertebrate interactions in the soil–litter system. Arrows represent directions of transfers.

affected by combinations of cadmium–lead and cadmium–lead–zinc. In another isopod, *Porcellio scaber*, Beeby (1978) studied calcium and lead uptake but made no conclusions concerning their effects on mortality. However, the fecundity of aphids fed tomatoes contaminated with mercury is decreased (Haney and Lipsey, 1973).

Heavy metals exert an effect on litter decomposition processes in terrestrial ecosystems and there is considerable evidence that metals can be responsible for a decreased rate of decomposition concomitant with an increase in the total litter biomass. It is likely that this action is via an effect on microorganism diversity and activity subsequently affecting the production of essential enzymes. Such an effect is complicated by a further effect on litter-inhabiting organisms which have been involved in complex interactions with microorganisms, both in terms of the outside environment (Addison and Parkinson, 1978; Ausmus, 1977), the internal environment of the animal gut (e.g. the case of the calcium cycle, Cromack *et al.*, 1977) and in animal excretions (Webb, 1977). It is thus surprising that this part of the litter ecosystem has been so little studied until recently, although any such study would require knowledge of the complexity of the processes involved, some of which are summarized in Fig. 4.4.

Quite apart from the disruption of these processes by heavy metals, there also remain the problems of litter accumulation and food chain transference of metals. It is clear that food chain transference of some heavy metals does take place (see Section 3) and considering the importance of both microorganisms and invertebrates in litter decomposition processes the lack of relevant data (field or experimental) concerning the effects of heavy metals represents one area where a great deal of further research is required. Nutrient turnover in forest ecosystems is well documented (e.g. Lang, 1978; MacLean and Wien, 1978) but apart from the work of van Hook *et al.* (1977, 1978) few attempts have been made to associate nutrient cycles with those of heavy metals. Until an integrated source of data is available it remains difficult to assess the long-term implications of heavy metal pollution on terrestrial ecosystems.

5. LONG-TERM EFFECTS OF HEAVY METALS IN ECOSYSTEMS

Features that should be stressed when reviewing and assessing the impact of heavy metal pollution on ecosystems are that heavy metals often occur together, zinc and cadmium being a classic example (although zinc and lead also commonly occur together). Pollution by heavy metals is also frequently part of a 'suite' of pollutants usually including oxides of sulphur. On metal mining sites, etc., it is common to find that the occurrence of high concentrations of metals is also accompanied by low levels of major plant nutrients. We are therefore faced with a situation in the field where the separation and identification of effects which can be attributed solely to pollution by heavy metals becomes difficult.

Figure 4.5 demonstrates both immediate and long-term effects of heavy metals on terrestrial ecosystems. Each part or process can be considered as a dynamic equilibrium, a change in which will effect a change in another process. Initial effects of aerial pollution can be considered in terms of the litter ecosystem solely while the soil remains relatively uncontaminated. In contrast, effects of parent material weathering are on an already contaminated soil system. Naturally, the soil and litter systems are themselves closely interlinked and transfers between the components are complex (see Fig. 4.1).

In considering the effects of metals on ecosystems it is important to recognize that the impact will depend on the type of ecosystem under consideration. In the simplest case there will be two distinct situations which will result in different ecosystem responses, and the distinction can be made on the mode of imposition of the heavy metal stress.

In the first situation, massive contamination occurs in the substrate resulting from either natural outcropping or exposure of geological strata rich in heavy metal ores, or from the dumping of mining/smelting wastes in which appreciable amounts of heavy metals remain. In these situations there are usually severely toxic concentrations of heavy metals in the soil, which becomes colonized by species or ecotypes able to tolerate such conditions. Often the process of colonization is extremely slow, but over a long period of time a successional development of plant communities adapted to high levels of heavy metals can be expected. Ecosystems developed in this way progressively increase in complexity provided that no other intervention takes place. Examples are the colonization of mining spoil by metal-tolerant plant species (Antonovics *et al.*, 1971; Bradshaw, 1976; Gemmel, 1978; Wild, 1974*b*) and the occurrence of distinctive vegetation on natural outcropping heavy metal substrates. In this latter case the vegetation of Serpentine soils (containing nickel, chromium and cobalt) and copper-rich soils provide examples of more complex ecosystems, which, although stable in these contaminated situations nevertheless support physiognomically very different ecosystems to nearby areas where heavy metals are present in normal background levels

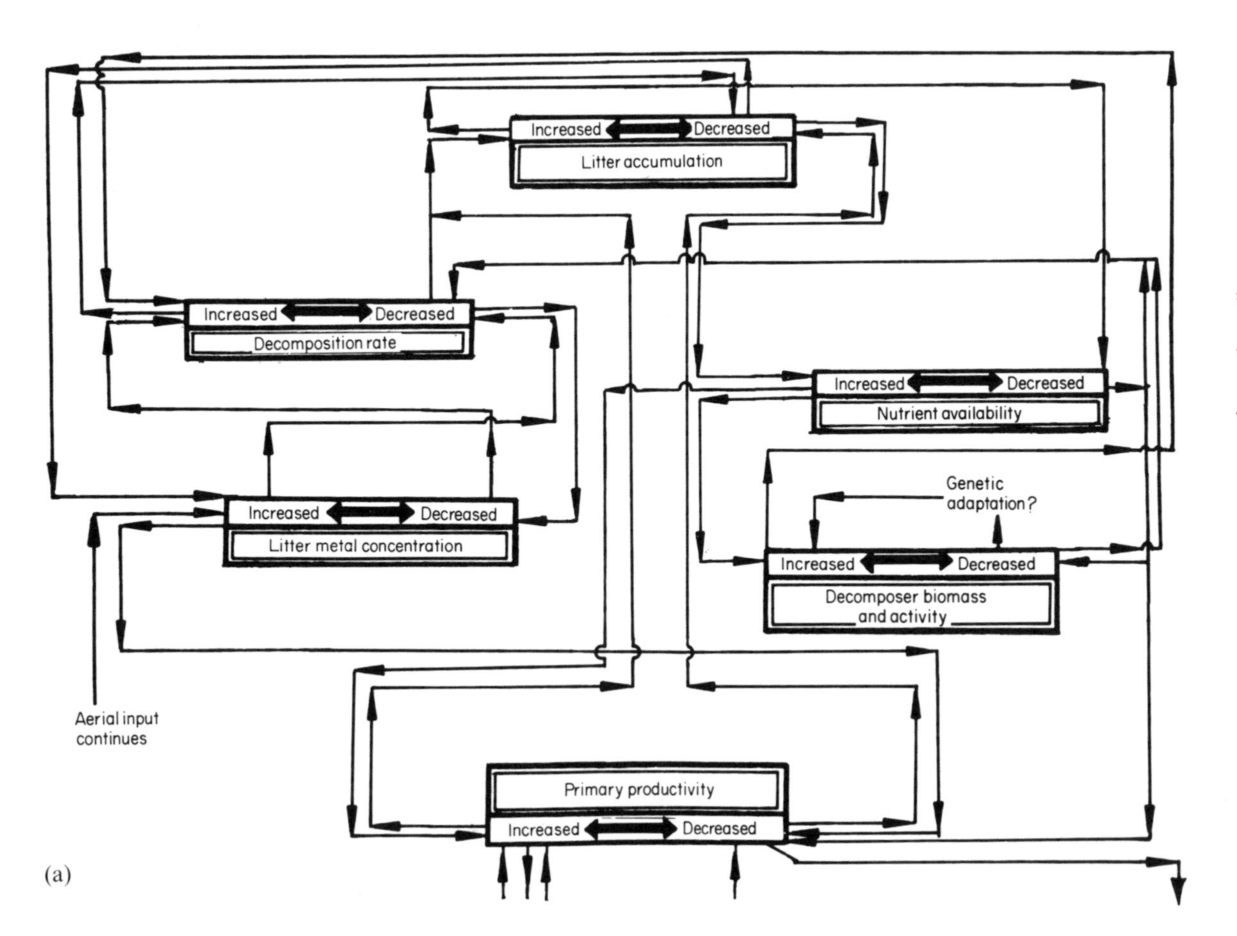
Increased
Decreased
Litter accumulation
Increased
Decreased
Decomposition rate
Increased
Decreased
Nutrient availability
Increased
Decreased
Litter metal concentration
Genetic adaptation?
Increased
Decreased
Decomposer biomass and activity
Aerial input continues
Primary productivity
Increased
Decreased

(a)

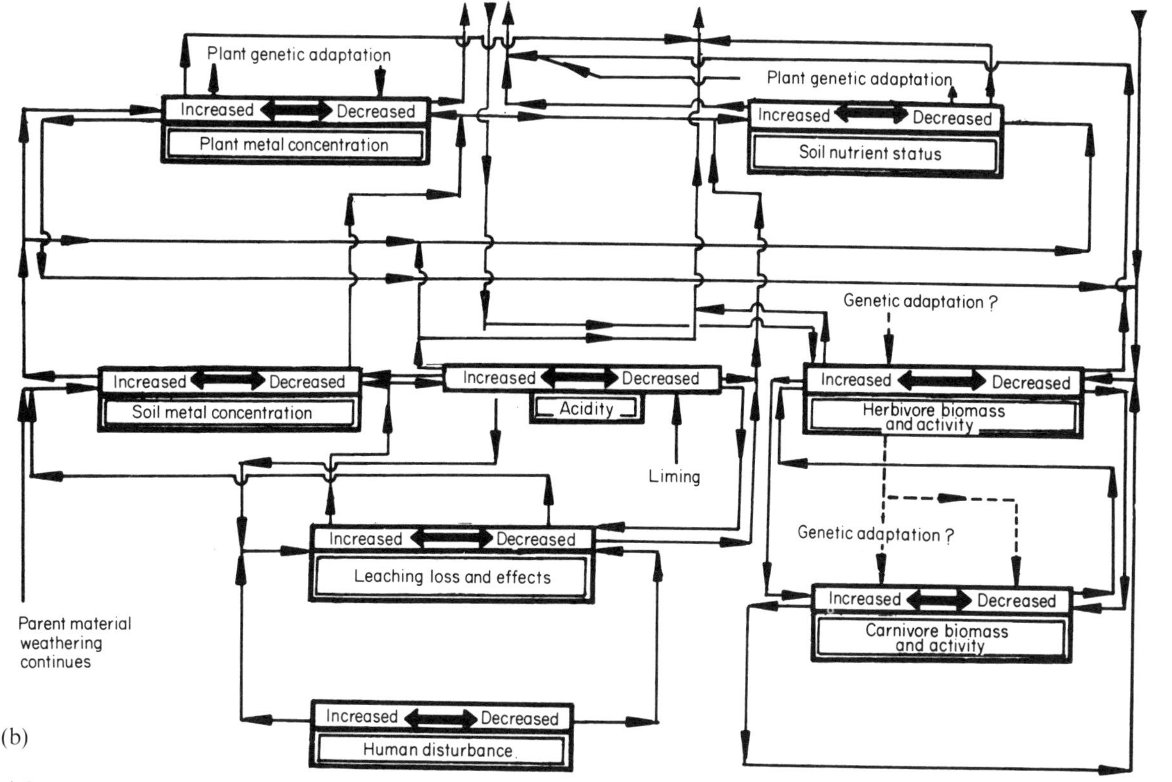

FIG. 4.5 A summary of heavy metal effects on terrestrial ecosystems. Each process and effect is represented as an equilibrium which may be increased or decreased according to the influence of another process or effect.

(Bølviken and Lag, 1977; Brooks, 1972; Ernst, 1974; Henwood, 1857; Malaise *et al.*, 1979; Rune, 1953; Wild, 1968, 1974*a*).

In the second case, contamination occurs by aerial deposition, continued application of sewage sludges containing heavy metals or other similar processes in which the ecosystem receives a continued input of heavy metals. In the majority of instances the impact of such progressive, and often gradual, build up of heavy metal concentrations is on previously established, relatively stable and complex ecosystems. Once a threshold amount of metal for the most sensitive components and processes in the ecosystem is reached, the sequence of events proposed by Holdgate (1979) can be expected to occur (Section 1). If the contamination is sufficiently severe or long term, a process which May (1976) has likened to a reversal of successional patterns may take place. In comparison to mining sites, plant tolerance in aerially contaminated sites has received little attention. Simon (1977) and Coughtrey and Martin (1977*b*, 1978*a*) reported data for cadmium tolerance in plant species from such sites. However, there is increasing evidence (Coughtrey *et al.*, 1978) that such tolerance does not follow the patterns proposed by Antonovics *et al.* (1971) for plants from mining sites. This suggestion is supported by the recent work of Cox and Hutchinson (1979) where the co-tolerance of *Deschampsia caespitosa* to metals other than those contaminating the parent soils was discussed. Furthermore, Coughtrey and Martin (1978*b*) provide evidence to suggest that *Holcus lanatus* collected from sites aerially contaminated by fallout from a lead/zinc smelter shows multiple tolerance to lead, zinc and cadmium. Essentially, aerially contaminated sites might be expected to differ from the other contaminated sites, discussed earlier, by the following factors:

(1) A lesser degree of contamination (although continuous).
(2) A lack of major disruption, at least in the first instance, of soil nutrient conditions.
(3) No physical destruction of the soil aggregates.
(4) The lack of clear-cut boundaries between contaminated and uncontaminated sites.

It seems likely that this combination of factors may well operate adaptive mechanisms which differ from those of mining or naturally contaminated sites and thus create other complications in the furtherance of conclusions concerning the long-term ecosystem effects of heavy metals. An example would be the problems arising from revegetation of mine wastes or spoils using plant species selected for metal tolerances; such plants contain high concentrations of heavy metals which are then directly available to higher trophic levels.

Although the effects of genetic adaptation are included in the processes summarized in Fig. 4.5, it should be made quite clear that adaptation of one component alone does not protect the ecosystem as a whole, it serves only to alter one dynamic equilibrium, and can result in a series of actions on other equilibria represented, some of which may be advantageous, and some detrimental, to the ecosystem as a whole. It is highly unlikely that one set of adaptations would prevent a change in the ecosystem structure and diversity.

6. CONCLUSIONS

It is clear, at the time of writing, that heavy metals do exert effects on terrestrial ecosystems, and that these effects act mainly to the detriment of the ecosystem (see Fig. 4.5). It is also apparent that the projected effects will differ according to the mode and degree of contamination. There is considerable laboratory evidence to support the effects discussed, but relatively little field evidence. The overriding problem of heavy metal contamination is long-term persistence. Apart from the work of Roberts and Goodman (1974) there are no field data concerning ecosystem change subsequent to the cessation of an aerial input. However, Tyler (1978) calculated the 10% residence times for heavy metals in soils and suggested a range of 3 years (manganese) to 90 years (lead) in unpolluted soils, but from 2 years (vanadium) to more than 200 years (lead) in polluted soils. These data were obtained by leaching soils with precipitation water at pH 4·2, which is considerably lower than expected, assuming that the cessation of aerial deposition of metals would be associated with a decrease in atmospheric nitrogen and sulphur levels following closure of an industrial site.

The insidious nature of aerial contamination and its dissemination within the environment and specific ecosystems has been discussed and it is apparent that such contamination, whether resulting from industrial or agricultural operations, cannot immediately be prevented. A much greater knowledge of the biological effects of heavy metals with reference to field studies is required if any firm conclusions are to be drawn and hence any proposals forthcoming. Studies of heavy metals in the environment and on ecosystems in particular are complicated by the extremely complex interaction of physical, chemical and biotic factors. Although there is evidence for adaptation (both physiological and genetic) within individual components of ecosystems, this should not be considered as a 'cure' for ecosystem effects since adaptation at only one stage may lead to irreversible effects at another stage. The problems of irreversible effects either within an ecosystem, or on a whole ecosystem, should be considered during assessment procedures for heavy metal release. The highest trophic level of many terrestrial ecosystems is man, the progressive accumulation of metals within these ecosystems could be viewed as a threat to his well being, either manifest directly, as increased metal concentrations of his foodstuffs, or indirectly, in terms of a degradation in the quality of his environment.

ACKNOWLEDGEMENT

We wish to thank the Natural Environment Research Council for funding during the period when this article was written.

REFERENCES

Addison, J. A. and D. Parkinson (1978). Influence of collembolan feeding activities on soil metabolism at a high arctic site. *Oikos*. **30**: 529–38.

Alderman, G. A. (1968). Some aspects of soil–plant–animal relationships. In: *Trace Elements*. Welsh Soils Discussion Group Report No. 9. (P. Jenkins (ed)), pp. 109–19.

Allaway, W. H. (1968). Agronomic controls over the environmental cycling of trace elements. *Adv. Agron.* **20**: 235–74.

Allcroft R. (1951). Lead poisoning in cattle and sheep. *Vet. Rec.* **63**: 583–90.

Andersson, A. (1976). On the determination of ecologically significant fractions of some heavy metals in soils. *Swed. J. Agric. Res.* **6**: 19–25.

Andersson, A. (1977). Heavy metals in Swedish soils: On their retention, distribution and amounts. *Swed. J. Agric. Res.* **7**: 7–20.

Andren, A. W., J. A. C. Fortescue, G. S. Henderson and D. E. Reichle (1973). Environmental monitoring of toxic metals in ecosystems. In: *Ecology and Analysis of Trace Contaminants*. ORNL-NSF-EATC-1, Oak Ridge National Laboratory, Tennessee, pp. 61–119.

Antonovics, J., A. D. Bradshaw and R. G. Turner (1971). Heavy metal tolerance in plants. *Adv. Ecol. Res.* **7**: 1–85.

Arvik, W. H. and R. L. Zimdahl (1974). Barriers to the foliar uptake of lead. *J. Environ. Qual.* **4**: 369–73.

Ausmus, B. S. (1977). Regulation of wood decomposition rates by arthropod and annelid populations. *Ecol. Bull. (Stockholm)*. **25**: 180–92.

Ausmus, B. S., D. R. Jackson and G. J. Dodson (1977). Assessment of microbial effects on ^{109}Cd movement through soil columns. *Pedobiologia*. **17**: 183–8.

Ausmus, B. S., G. J. Dodson and D. R. Jackson (1978). Behaviour of heavy metals in forest microcosms. III. Effects on litter–soil carbon metabolism. *Water Air Soil Pollut.* **10**: 19–26.

Babich, H. and G. Stotsky (1977). Sensitivity of various bacteria, including actinomycetes, and fungi to cadmium and the influence of pH on sensitivity. *Appl. Environ. Microbiol.* **33**: 681–95.

Balicka, N., T. Wegryzn and E. Czekanowska (1977). Microorganisms as indices of environmental pollution by smelting industry. *Acta Microbiol. Pol.* **26**: 300–8.

Beeby, A. (1978). Interaction of lead and cadmium uptake by the woodlouse, *Porcellio scaber* (Isopoda, Porcellionidae). *Oecologia, Berl.* **32**: 255–62.

Blaylock, B. G., R. A. Goldstein, J. A. Huckabee, S. Jansen, D. Matti, R. G. Olmstead, M. Slawsley, R. A. Stella and J. P. Witherspoon (1973). Ecology of toxic metals. In: *Ecology and Analysis of Trace Contaminants.* ORNL-NSF-EATC-1, Oak Ridge National Laboratory, Tennessee, pp. 121–60.

Bolton, J. (1975). Liming effects on the toxicity to perennial ryegrass of a sewage sludge amended with Zn, Ni, Cu and Cr. *Environ. Pollut.* **9**: 295–304.

Bølviken, B. and J. Lag (1977). Natural heavy metal poisoning of soils and vegetation: An exploration tool in glaciated terrain. *Appl. Earth Sci.* **86**: 173–80.

Bond, H., B. Lighthart, R. Shimbaku and L. Russell (1976). Some effects of cadmium on forest soil and litter microcosms. *Soil Sci.* **121**: 278–87.

Bordeau, P. and M. Treshow (1978). Ecosystem response to pollution. Ch. 15 in: *Principles of Ecotoxicology* (G. C. Butler (ed)), John Wiley and Sons, Chichester.

Borg, K., H. Wanntrop, K. Erne and E. Hanko (1969). Alkyl mercury poisoning in terrestrial Swedish wildlife. *Vitrevy.* **6**: 301–76.

Bradshaw, A. D. (1976). Pollution and evolution. In: *Effects of Air Pollutants on Plants* (T. A. Mansfield (ed)), Cambridge University Press, Cambridge, pp. 135–60.

Brooks, R. R. (1972). *Geobotany and Biogeochemistry in Mineral Exploration.* Harper and Row, London. 290 pp.

Bull, K. R., R. D. Roberts, M. J. Inskip and G. T. Goodman (1977). Mercury concentrations in soil, grass, earthworms and small mammals near an industrial emission source. *Environ. Pollut.* **12**: 135–40.

Cenzi, G. and G. Morozzi (1977). Evaluation of the toxic effect of Cd^{2+} and $Cd(CN)_4^{2-}$ ions on the growth of mixed microbial populations of activated sludge. *Sci. Total Environ.* **7**: 131–43.

Cha, J. W. and B. W. Kim (1975). Ecological studies of plants for the control of environmental pollution. IV. Growth of various plant species as influenced by soil-applied cadmium. *Kor. J. Bot.* **18**: 23–30.

Chamberlain, A. C. (1975). The movement of particles in plant communities. In: *Vegetation and the Atmosphere. 1. Principles* (J. L. Monteith (ed)), Academic Press, London, pp. 155–203.

Chaney, R. L. (1973). Crop and food-chain effects of toxic elements in sludges and effluents. In: *Recycling Municipal Sludges and Effluents on Land,* US Environmental Protection Agency, Washington, pp. 129–41.

Chaney, W. R., J. M. Kelly and R. C. Strickland (1978). Influence of cadmium and zinc on carbon dioxide evolution from litter and soil from a Black Oak forest. *J. Environ. Qual.* **7**: 115–19.

Cole, M. A. (1977). Lead inhibition of enzyme synthesis in soil. *Appl. Environ. Microbiol.* **33**: 262–8.

Cole, M. A. and A. J. Turgeon (1978). Microbial activity in soil and litter underlying bandane and calcium arsenate treated turfgrass. *Soil Biol. Biochem.* **10**: 181–6.

Coughtrey, P. J. (1978). Cadmium in terrestrial ecosystems: a case study at Avonmouth, Bristol, England. Ph.D. Thesis, University of Bristol, Bristol.

Coughtrey, P. J. and M. H. Martin (1976). The distribution of Pb, Zn, Cd and Cu within the pulmonate mollusc, *Helix aspersa* Müller. *Oecologia, Berl.* **23**: 315–22.

Coughtrey, P. J. and M. H. Martin (1977*a*). The uptake of Pb, Zn, Cd and Cu by the pulmonate mollusc, *Helix aspersa* Müller, and its relevance to the monitoring of heavy metal contamination of the environment. *Oecologia, Berl.* **27**: 65–74.

Coughtrey, P. J. and M. H. Martin (1977*b*). Cadmium tolerance of *Holcus lanatus* from a site contaminated by aerial fallout. *New Phytol.* **79**: 273–80.

Coughtrey, P. J. and M. H. Martin (1978*a*). Cadmium uptake in tolerant and non-tolerant populations of *Holcus lanatus* grown in solution culture. *Oikos.* **30**: 555–60.

Coughtrey, P. J. and M. H. Martin (1978*b*). Tolerance of *Holcus lanatus* to lead, zinc and cadmium in factorial combination. *New Phytol.* **81**: 147–54.

Coughtrey, P. J., M. H. Martin and E. W. Young (1977). The woodlouse *Oniscus asellus* as a monitor of environmental cadmium levels. *Chemosphere.* **6**: 827–32.

Coughtrey, P. J., M. H. Martin and S. W. Shales (1978). Preliminary observations on cadmium tolerance in *Holcus lanatus* L. from soils artificially contaminated with heavy metals. *Chemosphere.* **7**: 193–8.

Coughtrey, P. J., C. H. Jones, M. H. Martin and S. W. Shales (1979). Litter accumulation in woodlands contamined by Pb, Zn, Cd and Cu. *Oecologia, Berl.* **39**: 51–60.

Coughtrey, P. J., M. H. Martin, J. Chard and S. W. Shales (1980*a*). Micro-organisms and metal retention in the woodlouse *Oniscus asellus. Soil Biol. Biochem.* **12**: 23–7.

Coughtrey, P. J., M. H. Martin and S. W. Shales (1980*b*). A case study of the effects of airborne cadmium contamination on a terrestrial ecosystem. 2. Invertebrates and micro-organisms. In: *Case Studies in Ecotoxicology*. Surrey University Press, Guildford (in press).

Cox, R. M. and T. C. Hutchinson (1979). Metal co-tolerances in the grass *Deschampsia caespitosa. Nature, Lond.* **279**: 231–3.

Craig, R. B. and R. L. Rudd (1974). The ecosystem approach to toxic chemicals in the biosphere. In: *Survival in Toxic Environments* (M. A. Q. Khan and J. P. Bederka (eds)), Academic Press, New York, pp. 1–24.

Cromack, K., Jr, P. Sollins, R. L. Todd, D. A. Crossley, Jr, W. M. Fender, R. Fogel and A. W. Todd (1977). Soil microorganism–arthropod interactions: Fungi as a major calcium and sodium source. Chapter 9 in: *The Role of Arthropods in Forest Ecosystems* (W. J. Mattson (ed)), Springer-Verlag, Berlin, pp. 246–52.

Curnow, R. D., W. A. Tolin and D. W. Lynch (1977). Ecological and land use relationships of toxic metals in Ohio's terrestrial vertebrate fauna. In: *Biological Implications of Metals in the Environment*, Proceedings 15th Annual Hanford Life Sciences Symposium, Oak Ridge, Tennessee, ERDA-TIC-CONF-No. 750929, pp. 578–94.

De Leval, J. and J. De Monty (1972). Evolution de la microflore du sol en fonction de sa concentration en Zn et en Pb. *Rev. Ecol. Biol. Sol.* (1972): 491–504.
Denaeyer-De Smet, S. (1974). Premier apercu de la distribution du cadmium dans divers écosystèmes terrestres non pollués et pollués. *Oecol. Plant.* **9**: 169–82.
Denaeyer-De Smet, S. and P. Duvigneaud (1974). Accumulation de metaux lourdes dans divers écosystèmes terrestres pollués par des rétombées d'origine industrielle. *Bull. Soc. Roy. Bot. Belg.* **107**: 147–56.
Dick, A. T. and L. B. Bull (1945). Some preliminary observations on the effect of molybdenum on copper metabolism in herbivorous animals. *Aust. Vet. J.* **21**: 70–2.
Donovan, P. P., D. T. Feeley and P. P. Canavan (1969). Lead contamination of mining areas in W. Ireland. II. Survey of animals, pastures, foods and waters. *J. Sci. Food Agric.* **20**: 43–5.
Dorn, C. R., J. O. Pierce, G. R. Chase and P. E. Phillips (1975). Environmental contamination by lead, cadmium, zinc and copper in a new lead producing area. *Environ. Res.* **9**: 159–72.
Doyle, J. J., R. T. Marshall and W. H. Pfander (1975). Effects of cadmium on growth and uptake of cadmium by microorganisms. *Appl. Microbiol.* **29**: 562–4.
Draggan, S. (1976). The microcosm as a tool for estimation of environmental transport of toxic materials. *Int. J. Environ. Studies.* **10**: 65–70.
Ernst, W. (1974). *Schwermetallvegetation der Erde.* Fischer-Verlag, Stuttgart, 194 pp.
Ernst, W. (1976). Physiological and biochemical aspects of metal tolerance. In: *Effects of Air Pollutants on Plants* (T. A. Mansfield (ed)), Cambridge University Press, Cambridge, pp. 115–33.
Ferguson, W. S., A. H. Lewis and S. J. Watson (1940). The teart pastures of Somerset. Jeallots Hill Research Station Bulletin 1, Kynoch Press, Birmingham.
Florkowski, T., S. Piorek and M. M. A. Sassen (1979). An attempt to determine the tissue concentration of *Quercus robur* L. and *Pinus sylvestris* L. foliage by particulates from zinc and lead smelters. *Environ. Pollut.* **18**: 97–106.
Garten, C. T., Jr and R. C. Dahlman (1978). Plutonium in biota from an East Tennessee floodplain forest. *Health Phys.* **34**: 705–12
Garten, C. T., Jr, R. H. Gardner and R. C. Dahlman (1978). A compartment model of plutonium dynamics in a deciduous forest ecosystem. *Health Phys.* **34**: 611–19.
Gemmel, R. P. (1978). *Colonisation of Industrial Wasteland.* Edward Arnold, London, 75 pp.
Giles, F. E., S. G. Middleton and J. G. Grau (1973). Evidence for the accumulation of atmospheric lead by insects in areas of high traffic density. *Environ. Entomol.* **2**: 299–300.
Gingell, S. M., R. Campbell and M. H. Martin (1976). The effect of zinc, lead and cadmium pollution on the leaf surface microflora. *Environ. Pollut.* **11**: 25–37.
Gish, C. D. and R. E. Christiensen (1973). Cadmium, nickel, lead and zinc in earthworms from roadside soils. *Environ. Sci. Technol.* **7**: 1060–2.
Gracey, J. F. and J. R. Todd (1960). Chronic copper poisoning in sheep following the use of copper sulphate as a molluscicide. *Brit. Vet. J.* **116**: 405–8.
Griffiths, J. J. (1919). Influence of mines upon land and livestock in Cardiganshire. *J. Agric. Sci.* **9**: 366–95.

Griffiths, J. R. and G. A. Wadsworth (1980). Heavy metal pollution of farms near an industrial complex. In: *Inorganic Pollution and Agriculture*, Paper 6, Ministry of Agriculture, Fisheries and Food Reference Book 326, HMSO, London, pp. 70–6.

Gullino, M. and N. Fiusello (1976). Azione del piombo sui funghi. *Mic. Ital.* **2**: 27–32.

Haghiri, F. (1974). Plant uptake of cadmium as influenced by cation exchange capacity, organic matter, zinc and soil temperature. *J. Environ. Qual.* **3**: 180–2.

Hallsworth, E. G., S. B. Wilson and E. A. Greenwood (1960). Copper and cobalt in nitrogen fixation. *Nature, Lond.* **187**: 79–80.

Haney, A. and R. L. Lipsey (1973). Accumulation and effects of methyl mercury hydroxide in a terrestrial food chain under laboratory conditions. *Environ. Pollut.* **5**: 305–16.

Hartman, L. M. (1974). A preliminary report: Fungal flora of the soil as conditioned by varying concentrations of heavy metals. *Am. J. Bot.* **61** (Abstract Suppl.): 23.

Healey, W. B. (1968). Ingestion of soil by dairy cattle. *N.Z.J. Agric. Res.* **11**: 487–99.

Healey, W. B. (1974). Ingested soil as a source of elements to grazing animals. In: *Trace Element Metabolism in Animals* (W. G. Hoekstra, J. W. Suttie, H. E. Ganther and W. Mertz (eds)), University Park Press, Baltimore, pp. 448–9.

Heinrichs, H. and R. Mayer (1977). Distribution and cycling of major and trace elements in two central European forest ecosystems. *J. Environ. Qual.* **6**: 402–7.

Henwood, J. (1857). Notice of the copper turf of Merioneth. *Edinburgh New Phil. J.* **3**: 61–3.

Hogl, O. (1975). Molybdan als toxischer faktor in einem Schweizer Alpental. *Mitt. Gebeite Lebensm. Hyg.* **66**: 485–95.

Holdgate, M. W. (1979). *A Perspective of Environmental Pollution*. Cambridge University Press, Cambridge, 278 pp.

Huckabee, J. W. and B. G. Blaylock (1973). The transfer of mercury and cadmium from terrestrial to aquatic ecosystems. In: *Metal Ions in Biological Systems—Studies of Some Biochemical and Environmental Problems* (S. K. Dhar (ed)), Plenum Press, London, pp. 125–60.

Hutchinson, T. C. and L. M. Whitby (1977). The effects of acid rainfall and heavy metal particulates on a boreal forest ecosystem near the Sudbury smelting region of Canada. *Water Air Soil Pollut.* **7**: 421–38.

Imai, K., T. Sugio, T. Tsuchika and T. Btano (1975). Effects of heavy metal ions on the growth and iron oxidising activity of *Thiobacillus ferroxidans*. *Agric. Biol. Chem.* **7**: 1349–54.

Inman, J. C. and G. R. Parker (1978). Decomposition and heavy metal dynamics of forest litter in Northwestern Indiana. *Environ. Pollut.* **17**: 39–51.

Ireland, M. P. (1975). Metal content of *Dendrobaena rubida* (Oligochaeta) in a base metal mining area. *Oikos*. **26**: 74–9.

Ireland, M. P. (1977). Lead retention in toads *Xenopus laevis* fed increasing levels of lead-contaminated earthworms. *Environ. Pollut.* **12**: 85–92.

Jackson, D. R., W. J. Selvidge and B. S. Ausmus (1978*a*). Behaviour of heavy metals in forest microcosms. 1. Transport and distribution among components. *Water Air Soil Pollut.* **10**: 3–11.

Jackson, D. R., W. J. Selvidge and B. S. Ausmus (1978*b*). Behaviour of heavy metals in forest microcosms. 2. Effects on nutrient cycling processes. *Water Air Soil Pollut.* **10**: 13–18.

Jarvis, S. C., L. H. P. Jones and M. J. Hopper (1976). Cadmium uptake from solution by plants and its transport from roots to shoots. *Plant Soil.* **44**: 179–91.

John, M. K. and C. van Laerhoven (1972). Lead uptake by lettuce and oats as affected by lime, nitrogen and sources of lead. *J. Environ. Qual.* **1**: 169–71.

Johnson, M. S. and R. D. Roberts (1978). Distribution of lead, zinc and cadmium in small mammals from polluted environments. *Oikos.* **30**: 153–9.

Joose, E. N. G. and J. B. Buker (1979). Uptake and excretion of lead by litter-dwelling Collembola. *Environ. Pollut.* **18**: 235–40.

Jordan, M. J. (1975). Effect of zinc smelter emissions and fire on chestnut–oak woodlands. *Ecology.* **56**: 78–91.

Jordan, M. J. and M. P. Lechavalier (1975). Effects of zinc smelter emissions on forest soil microflora. *Can. J. Microbiol.* **21**: 1855–65.

Kelly, J. M. and G. S. Henderson (1978). Effects of nitrogen and phosphorus additions on deciduous litter decomposition. *Soil Sci. Soc. Am. J.* **42**: 972–6.

Kimura, Y. and V. L. Miller (1964). The degradation of organomercury fungicides in soil. *Agric. Food Chem.* **13**: 253–7.

Lagerwerff, J. V. (1967). Heavy metal contamination of soils. In: *Agriculture and the Quality of Our Environment* (N. C. Brady (ed)), American Association of Advance Science Publishers, pp. 343–64.

Lang, G. E. (1978). Detrital dynamics in a mature oak forest: Hutchinson Memorial Forest, New Jersey. *Ecology.* **59**: 580–95.

Lawrey, J. D. (1977). Soil fungal populations and soil respiration in habitats variously influenced by coal strip-mining. *Environ. Pollut.* **14**: 195–205.

Lawrey, J. D. (1978). Trace metal dynamics in decomposing leaf litter in habitats variously influenced by coal strip-mining. *Can. J. Bot.* **56**: 953–62.

Letunova, S. V. (1970). Geochemical ecology of soil microorganisms. In: *Trace Element Metabolism in Animals* (C. F. Mills (ed)), E. and S. Livingstone, Edinburgh, pp. 432–7.

Lipman, C. B. and P. S. Burgess (1914). The effect of copper, zinc, iron and lead salts on ammonification and nitrification in soils. *Univ. Calif. Publ. Agric. Sci.* **1**: 127–39.

Little, P. (1973). A study of heavy metal contamination of leaf surfaces. *Environ. Pollut.* **5**: 159–72.

Little, P. (1977). Deposition of 2·75, 5·0 and 8·5 μm particles to plant and soil surfaces. *Environ. Pollut.* **12**: 293–305.

Little, P. and R. D. Wiffen (1977). Emission and deposition of petrol engine exhaust Pb. 1. Deposition of exhaust Pb to plant and soil surfaces. *Atmos. Environ.* **11**: 437–47.

Lu, P. Y., R. L. Metcalf, R. Furman, R. Vogel and J. Hasset (1975). Model ecosystem studies of lead and cadmium and of urban sewage sludge containing these elements. *J. Environ. Qual.* **4**, 505–9.

MacLean, D. A. and R. W. Wien (1978). Litter production and forest floor nutrient dynamics in pine and hardwood stands of New Brunswick, Canada. *Holarctic Ecol.* **1**: 1–15.

Maga, J. A. and F. B. Hodges (1972). A joint study on lead contamination relative to horses deaths in southern Solano County, State of California, Air Resources Board, California, USA.

Mahler, R. J., F. T. Bingham and A. L. Page (1978). Cadmium-enriched sewage sludge application to acid and calcareous soils: Effect on yield and cadmium uptake by lettuce and chard. *J. Environ. Qual.* **7**: 274–80.

Malaise, F., J. Gregoire, R. S. Morrison, R. R. Brooks and R. D. Reeves (1979). Copper and cobalt in vegetation of Fungurume, Shaba Provice, Zaire. *Oikos*. **33**: 472–8.

Malaney, G. W., W. D. Sheets and R. Quillin (1959). Toxic effects of metallic ions on sewage microorganisms. *Sewage Ind. Wastes*. **31**: 1309–15.

Martin, M. H. and P. J. Coughtrey (1975). Preliminary observations on the levels of cadmium in a contaminated environment. *Chemosphere*. **4**: 155–60.

Martin, M. H. and P. J. Coughtrey (1976). Comparisons between the levels of lead, zinc and cadmium within a contaminated environment. *Chemosphere*. **5**: 15–20.

Martin, M. H., P. J. Coughtrey and E. W. Young (1976). Observations on the availability of Pb, Zn, Cd and Cu in woodland litter and the uptake of Pb, Zn and Cd by the woodlouse *Oniscus asellus*. *Chemosphere*. **5**: 313–18.

Martin, M. H., P. J. Coughtrey, S. W. Shales and P. Little (1980). Aspects of airborne cadmium contamination of soils and natural vegetation. In: *Inorganic Pollution and Agriculture*, Paper 5, Ministry of Agriculture, Fisheries and Food Reference·Book 326, HMSO, London, pp. 55–69.

May, R. M. (1976). Patterns in multispecies communities. In: *Theoretical Ecology—Principles and Applications* (R. M. Ray (ed)), Blackwell, Oxford, pp. 142–62.

May, R. M. (1978). Factors controlling the stability and breakdown of ecosystems. In: *The Breakdown and Restoration of Ecosystems* (M. W. Holdgate and M. J. Woodman (eds)), Plenum Press, London, pp. 11–25.

Mendelssohn, H. (1962). Mass destruction of bird life owing to secondary poisoning from insecticides and rodenticides. *Atlantic Nat.* **17**: 247–8.

Mendelssohn, H. (1972). Effect of toxic chemicals on bird life. The impact of pesticides on bird life in Israel. Int. Council Bird Preservation XI, Bulletin 1972, pp. 75–104.

Munshower, F. F. (1972). Cadmium compartmentalisation and cycling in a grassland ecosystem in the Deer Lodge Valley, Montana. Ph.D. Thesis, University of Montana.

Munshower, F. F. (1977). Cadmium accumulation in plants and animals of polluted and non-polluted grasslands. *J. Environ. Qual.* **6**: 411–13.

Nielson, R. L. (1951). Effect of soil minerals on earthworms. *N.Z.J. Agric.* **83**: 433–5.

Odum, E. P. (1975). Diversity as a function of energy flow. In: *Unifying Concepts in Ecology* (W. H. Van Dobben and R. H. Lowe McConnell (eds)), Junk, The Hague, pp. 11–14.

Page, A. L., F. T. Bingham and C. Nelson (1972). Cadmium absorption and growth of various plant species as influenced by solution cadmium concentration. *J. Environ. Qual.* **1**: 288–91.

Parker, G. R., W. W. McFee and J. M. Kelly (1978). Metal distributions in forested ecosystems in urban and rural Northwestern Indiana. *J. Environ. Qual.* **7**: 337–42.

Peterson, P. J. (1971). Unusual accumulations of elements by plants and animals. *Sci. Prog. Oxford.* **59**: 505–26.

Pettersson, O. (1976). Heavy-metal ion uptake by plants from chemical solutions with metal ion, plant species and growth period variations. *Plant Soil.* **45**: 445–9.

Pettersson, O. (1977). Differences in cadmium uptake between plant species and cultivars. *Swed. J. Agric. Res.* **7**: 21–4.

Pinkerton, A. and J. R. Simpson (1977). Root growth and heavy metal uptake by three graminaceous plants in differentially limed layers of an acid, minespoil-contaminated soil. *Environ. Pollut.* **14**: 159–68.

Premi, P. R. and A. H. Cornfield (1969). Effects of Cu, Zn and Cr on immobilisation and subsequent remobilisation of nitrogen during incubation of soil treated with sucrose. *Geoderma.* **3**: 233–7.

Price, P. W., B. J. Rathcke and D. A. Gentry (1974). Lead in terrestrial arthropods: Evidence for biological concentration. *Environ. Entomol.* **3**: 370–2.

Rascio, N. (1977). Metal accumulation by some plants growing on zinc mine deposits. *Oikos.* **29**: 250–3.

Regier, H. A. and E. B. Cowell (1972). Applications of the ecosystem theory, succession, diversity, stress and conservation. *Biol. Cons.* **4**: 83–8.

Roberts, R. and M. S. Johnson (1978). Dispersal of heavy metals from abandoned mine workings and their transference through terrestrial food chains. *Environ. Pollut.* **16**: 293–310.

Roberts, T. M. and G. T. Goodman (1974). The persistence of heavy metals in soils and natural vegetation following closure of a smelter. In: *Trace Substances in Environmental Health—VII* (D. D. Hemphill (ed)), University of Missouri, Columbia, pp. 105–16.

Roberts, K. R., W. J. Miller, P. E. Stake, R. P. Gentry and M. W. Neathery (1974). Effects of high dietary Cd on Zn absorption and metabolism in calves fed for comparable nitrogen balances. *Proc. Soc. Expl. Biol. Med.* **144**: 906–8.

Roberts, R. D., Johnson, M. S. and M. Hutton (1978). Lead contamination of small mammals from abandoned metalliferous mines. *Environ. Pollut.* **15**: 61–9.

Ruhling, A. and G. Tyler (1973). Heavy metal pollution and decomposition of spruce needle litter. *Oikos.* **24**: 402–16.

Rune, O. (1953). Plant life on Serpentines and related rocks in the north of Sweden. *Acta Phytogeographica Suecica.* **31**: 1–139.

Rutherford, G. K. and C. R. Bray (1979). Extent and distribution of soil heavy metal contamination near a nickel smelter at Coniston, Ontario. *J. Environ. Qual.* **8**: 219–22.

Sharma, R. P. and J. L. Shupe (1977*a*). Trace metals in ecosystems: Relationships of the residues of Cu, Mo, Se and Zn in animal tissues to those in vegetation and soil in the surrounding environment. In: *Biological Implications of Metals in the Environment.* Proceedings 15th Annual Hanford Life Sci. Symp., Oak Ridge National Lab., ERDA-TIC-CONF-no. 750929, Tennessee, pp. 595–609.

Sharma, R. P. and J. L. Shupe (1977*b*). Lead, cadmium and arsenic residues in animal tissues in relation to their surrounding habitat. *Sci. Total Environ.* **7**: 53–62.

Simon, E. (1977). Cadmium tolerance in populations of *Agrostis tenuis* and *Festuca ovina*. *Nature, Lond.* **265**: 328–30.

Smith, W. H. (1976). Lead contamination of the roadside ecosystem. *J. Air Pollut. Control. Assoc.* **26**: 753–66.

Sorteberg, A. (1974). The effect of some heavy metals in oats in a pot experiment with three different soil types. *J. Sci. Agric. Soc. Finland.* **46**: 277–88.

Southwood, T. R. E. (1976). Bionomic strategies and population parameters. In: *Theoretical Ecology* (R. M. May (ed)), Blackwell Scientific Publications, Oxford, pp. 26–48.

Spalding, B. P. (1979). Effects of divalent metal chlorides on respiration and extractable enzymatic activities of Douglas Fir needle litter. *J. Environ. Qual.* **8**: 105–9.

Stickel, W. H. (1975). Some effects of pollutants in terrestrial ecosystems. In: *Ecological Toxicological Research* (A. D. McIntyre and C. F. Mills (eds)), Plenum Press, New York, pp. 25–74.

Stokinger, H. E. (1963). Effects of air pollutants on wildlife. *Conn. Med.* **27**: 487–92.

Strojan, C. L. (1978*a*). Forest leaf litter decomposition in the vicinity of a zinc smelter. *Oecologia, Berl.* **32**: 203–12.

Strojan, C. L. (1978*b*). The impact of zinc smelter emissions on forest litter arthropods. *Oikos.* **31**: 41–6.

Tabatabai, M. A. (1977). Effects of trace elements on urease activity in soils. *Soil Biol. Biochem.* **9**: 9–13.

Tansley, A. G. (1935). The use and abuse of vegetational concepts and terms. *Ecology.* **16**: 284–307.

Tatsuyama, K., H. Egawa, H. Yamamoto and H. Senmaru (1975*a*). Cadmium resistant microorganisms in the soil polluted by the metals. *Trans. Mycol. Soc. Japan.* **16**: 68–78.

Tatsuyama, K., H. Egawa, H. Yamamoto and H. Senmaru (1975*b*). Tolerance of cadmium resistant microorganisms to the other metals. *Trans. Mycol. Soc. Japan.* **16**: 79–85.

Timoney, J. F., J. Port, J. Giles and J. Spanier (1978). Heavy metal and antibiotic resistance in the bacterial flora of sediments of New York Bight. *Appl. Environ. Microbiol.* **36**: 465–72.

Tonomura, K., K. Maeda, F. Futai, T. Nakagami and M. Yamada (1968). Stimulative vaporisation of phenyl mercuric acetate by mercury resistant bacteria. *Nature, Lond.* **217**: 644–6.

Tyler, G. (1970). Blyets fordelning i ett sydvenskt barrskogsekosystem. *Grundfobattring.* **23**: 45–9.

Tyler, G. (1972). Heavy metals pollute nature, may reduce productivity. *Ambio.* **1**: 52–9.

Tyler, G. (1975*a*). Effects of heavy metal pollution on decomposition in forest soils. 1. Introductory, investigations. National Swedish Environment Protection Board, Solna, 21 pp.

Tyler, G. (1975*b*). Effects of heavy metal pollution on decomposition in forest soils. 2. Decomposition rate, mineralisation of nitrogen and phosphorus, soil enzymatic activity. National Swedish Environment Protection Board, Solna, 47 pp.

Tyler, G. (1975*c*). Heavy metal pollution and mineralisation of nitrogen in forest soils. *Nature, Lond.* **255**: 701–2.

Tyler, G. (1976). Heavy metal pollution, phosphatase activity and mineralisation of organic phosphorus in forest soils. *Soil Biol. Biochem.* **8**: 327–32.

Tyler, G. (1978). Leaching rates of heavy metal ions in forest soil. *Water Air Soil Pollut.* **9**: 137–48.

Tyler, G., N. Mornsjo and B. Nilsson (1974). Effects of cadmium, lead and sodium salts on nitrification in mull soil. *Plant Soil.* **40**: 237–42.

Uchida, Y., A. Saito, H. Kaziwara and N. Enomoto (1973). Studies on cadmium resistant microorganisms. 1. Isolation of cadmium resistant bacteria and the uptake of cadmium by the organisms. *Agric. Bull. Saga Univ.* **35**: 15–24.

van Hook, R. I. (1974). Cadmium, lead and zinc distributions between earthworms and soils: Potentials for biological accumulation. *Bull. Environ. Contam. Toxicol.* **12**: 509–12.

van Hook, R. I. and A. J. Yates (1973). Transient behaviour of cadmium in a grassland arthropod foodchain. *Environ. Res.* **9**: 76–83.

van Hook, R. I., B. G. Blaylock, E. A. Bondietti, C. W. Francis, J. W. Huckabee, D. E. Reichle, F. H. Sweeton and J. P. Witherspoon (1976). Radioisotope techniques in the delineation of the environmental behaviour of cadmium. Environ. Qual. Saf. **5**: 166–82.

van Hook, R. I., W. F. Harris and G. S. Henderson (1977). Cd, Pb and Zn distributions and cycling in a mixed deciduous woodland. *Ambio.* **6**: 281–6.

van Hook, R. I., D. R. Jackson and A. D. Watson (1978). Cd, Pb, Zn and Cu distributions and effects in forested watersheds. In: *Cadmium 77. Edited Proceedings of the 1st International cadmium Conference, San Francisco*, Metal Bulletins Ltd, London, pp. 140–9.

van Hook, R. I., D. W. Johnson and B. P. Spalding (1980). Zinc distribution and cycling in forest ecosystems. Chapter 12 in: *Zinc in the Environment. Part 1. Ecological Cycling* (J. O. Nriagu (ed)), John Wiley and Sons Inc., New York, pp. 419–37.

van Rhee, J. A. (1975). Copper contamination effects on earthworms by disposal of pig wastes in pastures. In: *Progress in Soil Zoology* (J. Vanek (ed)), Junk, The Hague, pp. 451–7.

Varanka, M. W., Z. M. Zablocki and T. D. Hinesley (1976). The effect of digested sludge on soil biological activity. *J. Water Pollut. Control Fed.* **48**: 1728–40.

Varma, M. M., W. A. Thomas and C. Prasad (1976). Resistance to inorganic salts and antibiotics among sewage-borne Enterobacteriaceae and Achromobacteriaceae. *J. Appl. Bacteriol.* **41**: 347–9.

Vetter, V. H. and R. Mahlop (1971). Untersuchungen uber Blei-Zink-und Fluorimmissionen und dadurch verursachte Schaden an Pflanzen und Tieren. *Landwirtsch. Forsch.* **24**: 294–315.

Watson, A. P. (1975). Trace element impact on forest floor litter in the new lead belt region of southeastern Missouri. In: *Trace Elements in Environmental Health—IX* (D. D. Hemphill (ed)), University of Missouri, Columbia, pp. 227–36.

Webb, D. P. (1977). Regulation of deciduous forest litter decomposition by soil arthropod faeces. Chapter 7 in: *The Role of Arthropods in Forest Ecosystems* (W. J. Mattson (ed)), Springer-Verlag, Berlin.

White, D. H., J. R. Bean and J. R. Longcore (1977). Nationwide residues of mercury, lead, cadmium, arsenic and selenium in starlings 1973. *Pestic. Monit. J.* **11**: 35–9.

Wieser, W., R. Dallinger and G. Busch (1977). The flow of copper through a terrestrial food chain. 2. Factors influencing the copper content of isopods. *Oecologia, Berl.* **30**: 265–72.

Wild, H. (1968). Geobotanical anomalies in Rhodesia. 1. The vegetation of copper bearing soils. *Kirkia*. **7**: 1–71.

Wild, H. (1974*a*). Geochemical anomalies in Rhodesia. 4. The vegetation of arsenical soils. *Kirkia*. **9**: 243–64.

Wild, H. (1974*b*). Arsenic tolerant plant species established on arsenical mine dumps in Rhodesia. *Kirkia*. **9**: 265–78.

Williams, S. T., T. NcNeilly and E. M. H. Wellington (1977). The decomposition of vegetation growing on metal mine waste. *Soil Biol. Biochem.* **9**: 271–5.

Williamson, P. (1980). Variables affecting body burdens of lead, zinc and cadmium in a roadside population of the snail *Cepaea hortensis* Müller. *Oecologia, Berl.* **44**: 213–20.

Williamson, P. and P. R. Evans (1972). Lead: Levels in roadside invertebrates and small mammals. *Bull. Environ. Contam. Toxicol.* **8**: 280–8.

Willoughby, R. A., E. Macdonald, B. J. McSherry and G. Brown (1972). The interaction of toxic amounts of lead and zinc fed to young growing horses. *Vet. Record*. 1972: 382–3.

Wilson, D. O. (1977). Nitrification in three soils amended with zinc sulphate. *Soil Biol. Biochem.* **9**: 277–80.

Woodwell, G. M. (1970). Effects of pollution on the structure and physiology of ecosystems. *Science*. **168**: 429–33.

Yost, K. J. (1978). Some aspects of the environmental flow of cadmium in the United States. In: *Cadmium 77. Edited Proceedings of the 1st International Cadmium Conference, San Francisco*, Metal Bulletins Ltd, London, pp. 147–66.

Yost, K. J. (1979). Some aspects of the environmental flow of cadmium in the United States. In *Cadmium Toxicity* (J. H. Mennear (ed)), Dekker Inc, New York, pp. 181–206.

Zhulidov A. V. and V. M. Emets (1979). Accumulation of lead in the bodies of beetles in contaminated environments associated with automobile exhausts. *Dokl. Akad. Nauk. SSSR*. **6**: 1515–16.

CHAPTER 5

Trace Metals in Agriculture

J. Webber

formerly Ministry of Agriculture, Fisheries and Food, Government Buildings, Lawnswood, Leeds LS16 5PY, UK†

Agricultural production is affected by trace metals in a number of ways. Some soils may contain above-normal concentrations as a result of the type of parent material from which they are derived. Others may, for similar reasons or because the elements present occur in forms unavailable to plants, exhibit trace element deficiencies which have to be corrected by applications, either to the land or directly to the crop, of appropriate chemicals in order to achieve satisfactory yields. Both excesses and deficiencies of trace metals may, because crop composition is affected, influence animal production.

Some agricultural practices result in additions of trace metals to the soil which may affect crops. Elements like copper and arsenic which are added to some animal diets to promote growth or control disease may later be applied to the land in farmyard manure. Fertilizers may also contain trace metals as impurities which, in the long term, can result in significant additions to the soil. In the past, plant protection chemicals have often contained elements like copper, mercury, lead or arsenic and still do to a lesser extent, so that their use can also add to the amounts of these elements present in the soil.

In addition to these agricultural practices, industry is also responsible either directly or indirectly for much trace element contamination of agricultural land. In many areas, wastes from mining or the treatment of metal ores have been dumped and, while such land is often derelict or rough

† Present address: Hazeldene, Mill Lane, Bardsey, Leeds LS17 9AN.

grazing, it may be in agricultural use—with consequent effects on crops or livestock. Direct emissions from industrial premises can cause local pollution by a range of trace elements, while a more widespread emission is that of lead in vehicle exhaust fumes as the result of additions normally made to petrol of organic lead compounds.

A further important route by which trace metals can reach agricultural land is through the use of waste products as manures or soil conditioners. The tanning industry, for example, produces large amounts of sludge which can be used as a source of slow-acting nitrogen for crops, but which contains appreciable amounts of chromium. Much of the metal waste from industry is discharged into the sewers, and, after treatment is concentrated in sewage sludge which may be applied to the land in large amounts. Trace metal pollution of soils from this source has received much consideration in recent years. Town refuse may also be composted, sometimes with the addition of sewage sludge, for use as a soil conditioner, and this too can add trace metals to the land.

Problems due to soil contamination with trace metals are essentially long term, since additions are cumulative and may only reach damaging proportions after many years. However, once soils have become contaminated they may remain in this state permanently, since removal of trace metals either through leaching by rainwater or by removal in crops is an extremely slow process. Various factors, notably soil reaction, can affect the availability to crops of metals in the soil; where heavy contamination has occurred, practices such as liming to raise the soil pH value and additions of organic matter can be used to ameliorate harmful effects on crops. The most important way of preventing toxic effects is to limit additions to the soil. To this end, guidelines for applications of metals in sewage sludge have been produced in many countries which could possibly be extended to some other forms of pollution.

1. NATURALLY OCCURRING TRACE METAL POLLUTION

In many mining districts soils are contaminated as a result of the presence of mineral lodes which may cause enrichment of adjacent areas (Davies, 1971), but it is difficult to separate this from contamination caused by mining and smelting activities. Problems arising from this source will be dealt with in the section on mining and related activities below (Section 3.1).

In several areas in Britain soils have been shown to contain molybdenum in amounts which can adversely affect livestock production. Herbage growing on some soils derived from Lower Lias shales has a molybdenum content so high as to cause molybdenosis (Le Riche, 1959) whilst on black Carboniferous shales in other areas, somewhat lower molybdenum concentrations can result in induced copper deficiency in cattle (Thompson *et al.*, 1972). Soils in some parts of Eire contain high concentrations of selenium and herbage growing on them can produce toxic effects in livestock (Fleming, 1962). Crop growth is not affected in any of these areas.

Crop growth is, however, affected by trace metals in some ultra-basic igneous rocks. Mitchell (1974) has reported nickel toxicity in cereals growing on soils with impeded drainage derived from such rocks in Scotland. Soils derived from Serpentine usually carry a limited range of plant species and are of poor agricultural value. Shewry and Peterson (1976) quote data from a number of workers who have reported nickel and chromium contents of such soils in the range 1500–5000 ppm with higher values from 2–12·5% chromium for samples from some tropical areas. Many of the plant species which can survive on such soils appear to accumulate nickel and/or chromium and to have developed a tolerance to high concentrations of these elements.

2. TRACE METAL ADDITIONS TO THE SOIL THROUGH AGRICULTURAL PRACTICES

2.1. Farmyard Manures from Animals Receiving Dietary Supplements

Both zinc and copper supplements may be added to pig diets, the former to correct possible deficiencies and the latter because it has been shown to improve the growth rate. Zinc is commonly added at the rate of about 100 ppm and copper at about 200–250 ppm, in both cases as the sulphate. A large proportion, usually over 80%, of the added trace metals is excreted by the pigs so that the resultant manure, most of which is used in crop production, contains high concentrations of copper and zinc. A number of workers (Berryman, 1970; Kornegay *et al.*, 1976; Unwin, 1980) have reported data on the composition of pig slurries and manures. The figures vary widely for individual samples but mean concentrations have mostly been in the range 675–1000 ppm for copper and 150–650 ppm for zinc when expressed on a dry matter basis. Kornegay *et al.* (1976) found for solid manure, from pigs with a rather high level of copper supplement (250–370 ppm) in the diet, that copper concentrations in the dry manure increased from 59–88 ppm when no supplement was given, to 1330–2367 ppm when copper was added.

The reported effects of these pig manures on crops are variable. Kornegay *et al.* (1976) found no significant increase in the copper content of maize grain after three years application at 13–15 t/ha although the copper content of the roots doubled from 5·6 to 11·2 ppm dry matter and there was a small increase in leaf copper concentration. Their results are given in Table 5.1. Gray and Biddlestone (1980) also found no increase in the copper contents of several vegetable crops following application of 120 kg/ha copper in a pig manure/straw compost. In both series of experiments the copper content of the soil was considerably increased.

TABLE 5.1
EFFECTS OF PIG MANURE ON COPPER AND ZINC CONTENTS OF MAIZE GRAIN (ppm). (AFTER KORNEGAY *et al.*, 1976)

Element	No manure		Low copper manure		High copper manure	
	2nd crop	3rd crop	2nd crop	3rd crop	2nd crop	3rd crop
Copper	1·43	2·79	1·19	2·46	1·60	3·17
Zinc	21·1	20·7	22·3	24·2	21·0	25·0

Most pig manure is applied to grassland. Batey *et al.* (1972) found that 12 kg/ha copper applied in three dressings of pig slurry increased the copper content of herbage, averaged over five cuts, from 9·1 to 21·2 ppm in the dry matter. They suggested that part of the increase was due to surface contamination with slurry. Similar results, demonstrating that average copper content of grass cut 28 days after slurry was applied increased from 14 to 50 ppm in the dry matter following applications containing a total of 47 kg/ha copper made over three years, were obtained by Gracey *et al.* (1976) who also thought that surface contamination from the summer dressings of slurry was partly responsible for the increase.

Both Gracey *et al.* (1976) and Kneale and Smith (1977) investigated the health of sheep grazed on slurry treated land. From serum enzyme (glutamic–oxalo acetic transaminase) determinations it was concluded that the animals were suffering severe liver breakdown due to excessive intake of copper. Kneale and Smith (1977) also found that the copper contents of the livers of lambs were raised from 25 to 112 ppm dry matter when grazed on grassland treated with pig slurry during the summer. In a second experiment, in which the slurry was subjected to anaerobic storage for 60 days before use, instead of being taken straight from the piggery, herbage copper contents were lower and concentrations in ewes livers, although high, were below those usually associated with toxicity. They postulated a change in the form of copper in the slurry during storage to one less available for uptake by grass.

There have been no reports of deaths of sheep on farms where copper-enriched pig manures were used, and Gracey *et al.* (1976) suggested that application rates up to 16 kg/ha copper annually could safely be used on grassland. This is higher than the maximum rate suggested for annual dressings on arable land of 9·3 kg/ha suggested by Chumbley (1971) for copper in sewage sludge.

Compounds of arsenic are commonly added to broiler rations and investigations on the manure show that the chemicals used, arsanilic acid, 4-nitrophenyl arsonic acid and 4-hydroxy-3-nitrophenyl arsonic acid, are, for the most part, excreted unchanged. Morrison (1969) investigated the effect of applying broiler litter on the arsenic contents of the soil and of crops

TABLE 5.2
ARSENIC CONTENTS OF SOILS AND CROPS FOLLOWING USE OF BROILER LITTER (ppm). (AFTER MORRISON, 1969)

No. of years applied	Soil	Crop dry matter	
		Lucerne	Clover
—	2·65	0·10	0·17
2	—	0·10	<0·10
20	1·83	0·12	0·15

grown on it. The litter used contained 15–30 ppm arsenic compared with <0·1 ppm and 2·6 ppm for two control samples. Analyses of the soil and of forage crops grown on a large broiler farm in Arkansas gave the results shown in Table 5.2. The dressings applied were those normally recommended for farm use, i.e. 10–15 t/ha adding 150–300 g/ha arsenic to the soil each year. If the litter analyses are typical there is little or no likelihood of arsenic toxicity to crops or danger to livestock following use of broiler litter from poultry receiving supplements of the element in their food. Bishop and Chisholm (1962) have suggested that a soil level of 50–125 ppm has to be reached before toxic effects appear even in sensitive crops.

2.2. Fertilizers

Modern fertilizer manufacturing processes result in products which contain only small quantities of impurities, so that while trace metals can be detected, the amounts present, especially in view of the low rates of application of fertilizers by comparison with those for bulky organic manures, rarely, if ever, present a hazard. Basic slag, a by-product of steel making, has been widely used, mainly on grassland, in Britain as a source of phosphate. It contains a wide range of trace metals mostly in forms unavailable to plants and there have been no reports of harmful effects on crops.

Rock phosphates from some sources, especially the Pacific Islands such as Nauru, often contain cadmium, which may be carried through to superphosphates used in commercial fertilizers. There is at present world-wide concern about levels of cadmium in the human diet and workers in Australia and the US have investigated the possible contribution of phosphate fertilizers in applications to the soil to these levels. Some recent data on the cadmium contents of fertilizers are given in Table 5.3.

TABLE 5.3

SOME REPORTED CADMIUM CONCENTRATIONS IN FERTILIZERS (ppm)

Fertilizer	Cadmium	No. of samples	References
Rock phosphate *ex* Nauru and Christmas Island	31–90	6	Williams and David (1973)
Fertilizers, 9–53% P_2O_5	18–91	21	Williams and David (1973)
US fertilizers, 10–46% P_2O_5	1·5–9·1	21	Lee and Keeney (1975)
Swedish fertilizers	0–30		Stenström and Vahter (1974)

Williams and David (1973) showed that for Australian fertilizers the cadmium and phosphate contents were closely correlated ($r = 0{\cdot}95$) and there was also a similar close relationship between the zinc and cadmium content, the mean zinc:cadmium ratio being 10:1, similar to that in rock phosphates. Lee and Keeney (1975) failed to find any such relationships for the Wisconsin fertilizers which they investigated, due, probably, to the wider range of raw materials and manufacturing processes used in the United States. Williams and David (1973) showed a considerable increase in cadmium content of topsoil following continued heavy use of super-phosphate. The cadmium was held in cation-exchangeable form and appeared to be readily available to crops. Some of their data from pot experiments comparing additions of cadmium as the chloride and in superphosphate are given in Table 5.4. Crops are known to vary considerably in the amounts of cadmium translocated to the parts normally eaten by man or animals and this is confirmed by these results, which also demonstrate the possibility of significant increases in dietary cadmium following use of fertilizers containing this element as an impurity. Much of the phosphate fertilizer used in Europe and America is derived from rocks with much lower cadmium contents than those used in Australia, and as shown in Table 5.3 contain much less cadmium. The risk to man from cadmium in fertilizers appears to be a very small one.

TABLE 5.4
CADMIUM CONTENTS OF EDIBLE PARTS OF VEGETABLE CROPS GROWN IN POTS (ppm DRY MATTER). (AFTER WILLIAMS AND DAVID, 1973)

Crop	Cadmium addition (μg/pot)		
	0	200 as cadmium chloride	85 as superphosphate
Peas	0·06	0·24	0·23
Beans	0·02	0·07	0·05
Tomato	0·07	0·34	0·16
Radish	0·14	1·14	0·66
Cabbage	0·08	0·41	0·22

2.3. Plant Protection Chemicals

In the past, copper-based fungicides were widely used in agriculture. For some perennial crops, notably vines and citrus fruits, where frequent spraying was practised, large amounts of the element could be added to the soil. Because the copper applied in this way accumulates mainly in the surface soil and the crops concerned are deep rooted, growth is not affected. If the vineyard or orchard is converted to growing other crops, copper toxicity problems can occur. The subject has been reviewed by Delas (1963). Since copper previously applied in this way is still likely to be present in toxic amounts, risks to shallow-rooted crops can persist for many years in the absence of recent applications. Copper fungicides were also used in the past for some annual crops, such as potatoes, but not in the same amounts or frequency as for vines or citrus fruits. No cases of copper toxicity in crops following such use have been reported. In recent years other fungicides have been developed, many of which do not contain trace metals, although there is some use of dithiocarbamates containing manganese and tin. There is no evidence of any harmful build-up of these elements in the soil.

Mercurial dressings for cereal crops are still widely used. Saha *et al.* (1970) investigated residues in cereal grains grown from seed which had been treated with 11·6–17 ppm mercury as methyl mercury dicyandiamide (I) or ethyl mercury *p*-toluene sulphonamide (II). Their results, given in Table 5.5, showed no noticeable increase in grain mercury as a result of the seed treatment. The concentrations found were in close agreement with

TABLE 5.5

MERCURY CONCENTRATIONS IN GRAIN GROWN FROM MERCURY-TREATED SEED (ppm). (AFTER SAHA *et al.*, 1970)

Crop	Control	(I) 17 ppm mercury	(II) 15·7 ppm mercury
Barley	0·010–0·014	0·008–0·015	0·010–0·012
Wheat	0·008–0·016	—	0·010–0·015

those reported by other workers. The small amounts of mercury added to the soil in seed dressings make it very unlikely that there could be any appreciable build-up of mercury concentrations. Work on grassland where relatively large amounts of mercurial fungicides had been applied on the greens of golf courses showed appreciable loss of the element by volatilization in 30–50 days (Kimura and Miller, 1964).

Other applications of mercury compounds in horticulture, as mercurous chloride to the soil, or as foliar sprays on fruit trees, also have negligible effects on amounts of the element found in the soil or crops. Concentrations in fruit, which should not be harvested less than six weeks after spraying, have not usually exceeded 0·070 ppm as compared with background levels of 0·005–0·040 ppm (Dept. Environment, 1976).

Lead arsenate sprays are used to control pests in apple orchards. Bishop and Chisholm (1962) reported arsenic concentrations ranging from 9·8 to 124·4 ppm, including 0·6–9·0 ppm in water-soluble forms, in samples taken from orchards in the Annapolis Valley, Canada. Soils in the same area on which lead arsenate had not been used had total arsenic contents from a trace level to 7·9 ppm. They suggested that more sensitive crops such as beans and strawberries, especially if grown on sandy soils, could suffer retarded growth where either total arsenic content was in the range 50–125 ppm or water-soluble arsenic above 2 ppm; it was also indicated that many of the orchard soils examined contained more than these amounts. In a later paper, Chisholm (1972) gave analytical data for a range of vegetable crops grown in soils containing up to 280 ppm lead and 25 ppm arsenic. A few samples contained more lead than the legally permitted level in foods offered for sale (2 ppm fresh weight), but arsenic contents were mostly in the range 0·1–0·25 ppm dry weight and the crops were quite safe in this respect for human consumption.

3. EFFECTS OF INDUSTRIAL CONTAMINATION OF SOILS

3.1. Mining and Related Activities

Considerable areas in western Britain have been affected by mining and treatment of metal ores, often for lengthy periods. Davies (1980) has reviewed the contamination arising as a result of these activities, dealing especially with the Devon/Cornwall area and Wales. Soil contamination in such areas can be owing to metal-bearing lodes occurring near the surface, to dumping of mine wastes, to dispersal of toxic materials by rivers (affecting floodplain soils in particular) and to atmospheric dispersal of pollutants from smelters. Land on which mine wastes have been tipped is often very heavily contaminated; Thornton (1980) reported concentrations of nearly 4 % lead near a lead 'rake' in Derbyshire and showed that values over 1000 ppm were usual within 500 m of the 'rakes', spoil heaps and smelter sites sampled. Some data for the trace metal contents of soils affected by mining are given in Table 5.6, illustrating the high concentrations which can occur. Recently, Anderson *et al.* (1979) have reported analyses for soil samples taken near Shipham in Somerset, with 1824 ppm lead, 7590 ppm zinc and 116 ppm cadmium, which are causing concern because of possible health hazards due to the unusually high cadmium concentrations found.

The agricultural problems found in these mining areas usually relate to the health of livestock. Stewart and Allcroft (1956) reported poor thriving of lambs in old lead mining areas in the Pennines and some deaths of stock also occur from time to time. The problems which occur have usually been associated with lead, and, since uptake of this element by plants even in contaminated areas is small, are thought in many cases to be associated with ingestion of soil. Thornton (1974) showed that animals in the Tamar valley could ingest 10 times as much copper, lead and arsenic in soil as that in the herbage they ate. Soil ingestion by cattle was found to range from 1·1 to 10·9 % of the total dry matter intake with mean values 4·5, 3·3 and 5·7 %

TABLE 5.6

METAL CONTENTS OF SOME SOILS AFFECTED BY MINING (ppm DRY SOIL). (AFTER DAVIES, 1976 AND THORNTON, 1980)

Metal	Davies (1976) 51 samples		Thornton (1980) 42 samples		'Uncontaminated' soil	
	Range	Mean	Range	Mean	Davies	Thornton
Lead	18–3 423	395	115–72 000	3 320	37	165
Zinc	44–1 905	232	41–1 260	247	94	128
Copper	3–523	56	10–204	46	13	48
Cadmium	0·4–2·3	1·2	<1–14	3	1·5	2
Mercury	0·01–1·78	0·25	—	—	0·03	—
Arsenic	—	—	23–1 080	228	—	—

Note: Davies's samples came from Devon/Cornwall and Wales as did those analysed for arsenic (23 samples) by Thornton. Thornton's samples came from Derbyshire and his 'uncontaminated' soils were from the same area and are not representative of Britain as a whole.

in November, January and March. These figures are in general agreement with those obtained by other workers.

The lead content of herbage grown in an uncontaminated soil was shown by Mitchell and Reith (1966) to rise from a few ppm during the period of active growth in summer to about 40 ppm in the dry matter for samples taken in autumn. Rather similar figures, given in Table 5.7, are reported by Thornton (1980) for samples taken from farms with different levels of lead pollution in Derbyshire, confirming the low level of uptake by plants. Arsenic, the other element which might, because of the poisonous nature of its compounds, be expected to cause livestock problems, has rarely been implicated in such cases. Arsenic uptake by crops is limited, as shown by

TABLE 5.7

MEAN LEAD CONTENTS OF PASTURE HERBAGE (ppm DRY MATTER) FROM THREE GROUPS OF DERBYSHIRE FARMS. (AFTER THORNTON, 1980)

Month of sampling	Low soil lead (100–200 ppm)	Medium soil lead (400–800 ppm)	High soil lead (1 000–2 000 ppm)
June	9	7	16
August	18	18	22
October	26	36	24

Thornton (1980) who reported herbage contents ranging from 0·26 to 9·60 ppm with mean 2·84 ppm for samples taken in April from crops growing on soils with 23–1080 ppm total arsenic. Porter and Peterson (1975) have, however, shown that some plant species can accumulate arsenic and may develop a tolerance to very high concentrations in the soil.

In Britain, much of the land affected by mining is under grass, so that effects on crop production are only rarely encountered. In some areas the zinc concentrations in the soil are so high that when pastures are ploughed and resown with grass severe yellowing and stunted growth occurs for a season or two. The problem tends to lessen with time although the reason for this is not known. Thomas (1980) has reported cases of reduced crop growth in Cornwall associated with high concentrations of arsenic in the soil but the areas involved are small. The effects of mining on crop production are evidently of only limited importance.

3.2. Atmospheric Emissions

The effects of atmospheric emissions of trace elements on soil concentrations and on crops have been investigated in many areas. In Britain, the area around Avonmouth near Bristol where a very large lead and zinc smelter is located has been extensively studied. Martin *et al.* (1980), from measurements of metal absorption by bags of *Sphagnum* moss, were able to produce maps of the levels of pollution in the area, showing that the amounts of trace metals deposited were related to wind direction. Griffiths and Wadsworth (1980) reported complaints by farmers of unthriftiness in cattle and showed that near the works soils and herbage had elevated concentrations of lead, zinc and cadmium (Table 5.8). Burkitt *et al.* (1972) reported that the cadmium content of ryegrass decreased from 50 ppm (dw) at 0·3 km from the works to 1·8 ppm (dw) at 11 km. There has been little or

TABLE 5.8

LEAD, ZINC AND CADMIUM CONTENTS OF SOILS AND HERBAGE NEAR A SMELTER AT AVONMOUTH. (AFTER GRIFFITHS AND WADSWORTH, 1980)

Distance from works	Total content of soil (ppm)·			Herbage content (ppm dry matter)		
	Lead	Zinc	Cadmium	Lead	Zinc	Cadmium
3 km NE	154	780	11	31	158	1·1
5 km NE	86	354	4	7	61	0·5
17 km SW	35	124	0·7	6	29	<0·25

Soil data based on 0–75 cm samples, results in ppm w/v of dried ground soil.

no evidence of crop growth being affected in the area, which may be partly due to the dilution of trace element accumulations on the soil surface by ploughing and cultivation. Investigations around smaller smelters in Yorkshire and Humberside have shown that contamination of soils and herbage was very localized and confined to fields adjacent to the works. Most of the emissions from smelters are in a particulate form, so that the pattern of deposition is, to a large extent, predictable from considerations of wind and local topography. There could be some risk of excessive accumulations of trace metals such as cadmium and lead in leafy vegetables if these were grown very close to the works.

A more widespread form of aerial pollution is that of lead from car exhausts. Lead added to petrol in organic lead compounds is emitted from car exhausts as particulate lead chloro-bromide (PbClBr) which may be deposited on grass and other crops growing near roads, especially those with high traffic densities. This has been widely investigated. Davies and Holmes (1972) confirmed that there were increased concentrations of lead in the soil and in grass near main roads but concluded that the amounts found in the herbage were below toxic levels for livestock. Lead concentrations fell off rapidly with distance from the road, and vegetables for human consumption could be safely grown at 50 m and beyond without risk to the consumer. Page and Ganje (1970) showed, however, that even at 1·5 km from a major road there could be some long-term accumulation of lead in the soil. Lagerwerff (1971) reported higher concentrations of lead,

TABLE 5.9

LEAD, ZINC AND CADMIUM CONTENTS OF SOILS AND RADISHES AS AFFECTED BY AERIAL CONTAMINATION. (AFTER LAGERWERFF, 1971)

Metal	Soil content (ppm)	Content of radishes (ppm dry matter)			
		Protected		Unprotected	
		Roots	Tops	Roots	Tops
Lead	30	19	13	21	26
	299	33	25	38	41
Cadmium	0·1	0·9	1·6	1·0	3·1
	0·6	1·7	2·9	1·8	5·0
Zinc	10	48	70	81	141
	60	82	152	121	248

Soils with lower metal concentrations taken 200 m from road, with higher concentrations 7 m from road. Crops grown in soil pH 5·9; protected under glass with filtered air, unprotected in the open at 200 m from road.

zinc and cadmium in soil taken 7 m from a busy road than at 200 m. Radishes were grown in soils from the two sampling points in pots situated 200 m from the road either exposed to the air or in a greenhouse with filtered air. Some of the results, given in Table 5.9, show that soil concentrations of all three metals were much higher at 7 m than at 200 m from the road. They also show that even at 200 m from the road, which carried 20 000 vehicles/24 h, aerial contamination accounted for 40 % or more of the metals in the radish tops. In the roots only zinc was significantly increased, due, it was thought, to translocation from the leaves.

Agricultural production is not affected to any significant extent by this source of pollution, much of which will fall on roadside verges, but in some situations excessive accumulations might occur in leafy vegetables. There would appear to be a good case for not growing such crops in any areas subject to atmospheric emissions of the more toxic trace metals such as lead, cadmium and arsenic, but the areas involved will in general be small.

4. USE OF INDUSTRIAL AND OTHER WASTES AS MANURES

Many industries produce wastes which contain trace metals either in the form of liquid effluent or as solid residues. Many liquid wastes are discharged into the sewers and the metals they contain are eventually concentrated in sewage sludge. Much of this goes to land; some more heavily contaminated sludges may be incinerated or dumped at sea, although in the latter case there are restrictions on the concentrations of some elements, such as cadmium and mercury, which they may contain. Solid wastes may be tipped on landfill sites, but where they are thought to have some value for agriculture, either as manures or soil conditioners, they may be spread on farm land. Waste sludge from tanneries is often used in this way. Town refuse is usually tipped or incinerated but sometimes, after screening to remove metal, glass, etc., it is composted, either alone or with the addition of sewage sludge, to produce an agricultural manure. Care needs to be taken in using all these materials to avoid excessive soil contamination and consequential damage to agriculture.

4.1. Tannery Waste

Wastes from tanneries are useful slow-acting sources of nitrogen for crops, although some of the nitrogen present may be contained in tanned hide which is very resistant to decomposition and solubilization of the nitrogen compounds present. The only recent reports on tannery wastes come from

TABLE 5.10

CHROMIUM CONTENTS OF SLUDGES FROM THREE TANNERIES IN POLAND (% DRY MATTER). (AFTER KOC *et al.*, 1976)

Tanning process	No. of samples	Range	Mean
Vegetable	17	0·78–1·90	0·97
Chrome	20	1·08–2·80	2·05
Vegetable/chrome	13	0·34–1·76	0·85

Poland (Koc *et al.*, 1976; Mazur and Koc, 1976*a*,*b*) where the effects on the soil and crops of sludges from three tanneries practising vegetable tanning, chrome tanning and a mixture of the two, were compared. The chromium contents of the dried sludges are given in Table 5.10.

In field experiments the sludges and farmyard manure were each applied annually at rates equivalent to 7·5 t/ha dry matter for a crop rotation of potatoes, spring barley, horse beans and winter wheat. Although there were differences between the sludges (with that from the vegetable tanning process giving better yields than the others and not significantly different from yields obtained with farmyard manure), there was no suggestion of chromium toxicity. There were, as expected, big increases in the chromium contents of the soils where the sludges had been applied but uptake by crops was small (Table 5.11). This is in agreement with other investigations of the effects of applied chromium on crops; these show that the element is

TABLE 5.11

CHROMIUM CONTENTS OF SOILS AFTER FOUR APPLICATIONS OF TANNERY SLUDGE AND FARMYARD MANURE AND OF CROPS GROWN OVER FOUR YEARS (ppm DRY MATTER). (AFTER MAZUR AND KOC, 1976*a*,*b*)

Tanning process	Soil chromium		Crops			
	Total	Extractable	Potato tubers	Barley grain	Horse beans	Wheat grain
Vegetable	175	7·5	1·0	3·3	1·0	1·4
Chrome	241	7·0	2·2	1·0	1·5	2·8
Vegetable/chrome	90	1·0	1·0	1·2	0·8	1·7
Farmyard manure	11·5	0·6	0·7	0·7	0·3	0·8

Extractable chromium soluble in 2·5% acetic acid at pH 2·5.

strongly adsorbed by the soil and so little translocated into the edible parts of crops that there has been concern as to whether some diets might not be deficient in the element. These experiments confirm practical experience with tannery sludges, showing them to be useful materials despite their often high chromium content. The absence of harmful effects and the small amounts taken up by crops is also in agreement with results obtained with sewage sludges with high chromium contents.

4.2. Composted Town Refuse

At different times, a number of Local Authorities in Britain have interested themselves in the composting of town refuse in order to produce a material which would be of value as a soil conditioner or manure. The commonest process has involved passing the refuse, following screening and grinding, with water or sewage sludge to moisten it, through slowly rotating drums, where, over a few days, fermentation starts. This is completed in large heaps outside, which are mechanically turned once or twice. In a few areas, notably Leicester, the process is still used and the compost is available to farmers and gardeners.

Experimental work on composted town refuse has been reported by Purves and Mackenzie (1973) and by Gray and Biddlestone (1980). The nutrient content of such composts is low but they contain considerable amounts of trace elements as shown in Table 5.12, which also gives data on the extractable trace element contents. Purves and Mackenzie (1973, 1974) found that effects on the yield of most of the vegetable crops which they tested varied with the season, but with dwarf beans there were significant reductions in growth with associated yellowing and necrosis of the leaves

TABLE 5.12

MEAN TOTAL AND EXTRACTABLE TRACE ELEMENT CONTENTS OF COMPOSTED TOWN REFUSE (ppm DRY MATTER). (AFTER GRAY AND BIDDLESTONE, 1980 AND PURVES, 1972)

Content	Copper	Zinc	Nickel	Cadmium	Lead	Boron
Total 1	610	1 350	140	7·5	1 630	174
Extractable 1	119	533	18	2·4	44	22
Extractable 2	136	468	14·5	—	44	11

Data: 1 after Gray and Biddlestone (1980), 2 after Purves (1972). For extractable contents 2·5% acetic acid was used except for copper for which Gray and Biddlestone used 0·05 M EDTA at pH 4·0 and for boron, for which hot water was used.

which they associated with boron toxicity. In their second paper (1974) they showed that the toxic effects following application of the compost, which contained 99 ppm water-extractable boron, were much reduced and yields greatly increased by leaching the compost with water, reducing the extractable boron content to 66 ppm.

Gray and Biddlestone (1980) reported the results of two field experiments, one on an alkaline clay soil and the other on an acid sandy soil, in which heavy dressings (about 250 t/ha/year) of composted town refuse were applied. As expected, trace element contents of the soils were increased, but the only harmful effect on the vegetable crops grown was a depression of seed germination and growth of dwarf beans. This was associated with large increases in the water-extractable boron contents of the compost-treated soils and was thought to be due to boron toxicity as found by Purves and Mackenzie (1973, 1974). Analyses of the vegetable crops, dwarf beans, lettuce, potatoes and spinach beet grown in the experiments showed some increases in trace element concentrations where the compost had been applied, but they were small in relation to the quantities added to the soil. For most crops composted town refuse can be used safely but, as for all materials adding to the trace metal burden of the soil, continued use on the same land may eventually produce toxic effects.

4.3. Sewage Sludge

A recent survey by the Standing Committee on the Disposal of Sewage Sludge (Dept. Environment, 1978) has shown that in the UK about 45% of the total dry solids in the sewage sludge produced is used on land as an aid to crop production. The amount used in this way was estimated to be about 550 000 t. A survey of the composition of sludges produced in England and Wales was made by Williams (1975) who showed that even when it came from sewage of purely domestic origin, sewage sludge contained appreciable amounts of trace metals. A more recent survey (Dept. Environment, 1980) of sludges going to land, including landfill, and representing 74% of total production, gave rather lower mean trace metal contents, which may be a better guide to the amounts of trace metals being added to agricultural soils. Some of the data is given in Table 5.13.

Other surveys of sludge composition have confirmed its variable nature and have shown that on occasion very high concentrations of particular elements can occur. Klein *et al.* (1974) investigated the sources of trace metals in New York City sewage and concluded that since 25–50% of the totals found appeared to come from domestic sources the possibility of greatly reducing concentrations in sludge by control of industrial effluents

TABLE 5.13

MEAN TRACE METAL CONTENTS OF SEWAGE SLUDGES FROM WORKS IN BRITAIN (ppm DRY MATTER)

Sludge	Zinc	Copper	Nickel	Lead	Chromium	Cadmium
England and Wales (all)	3 576	958	382	956	1 101	—
England and Wales (domestic)	1 320	30	40	390	70	—
Great Britain (to land)	1 820	613	188	550	744	29

Data for England and Wales from Williams (1975); for Great Britain from Dept. Environment (1981).

was much less than has often been argued. No similar published data appears to exist for other areas.

Trace metals in sewage sludge applied to land may affect agriculture by reducing crop growth or because increased amounts taken up by crops may render them unsuitable for human or animal food. For livestock, the ingestion of heavily contaminated soil could also present a hazard. An early survey of sludge composition by Berrow and Webber (1972) suggested that copper and zinc were the elements most likely to cause crop problems since they were usually present in concentrations much higher than those found in soils. Some sludges also contained nickel, an element highly toxic to plants, at levels which could affect crop growth. High concentrations of zinc and copper in the soil were also known (from advisory experience in England (Patterson, 1971) and from investigations by Rohde (1962) of the causes of 'exhaustion' of land irrigated with sewage at the Berlin and Paris sewage farms) to be implicated in reduced growth or even failure of crops following prolonged or very heavy use of sewage sludge.

The other trace metals in sewage sludge which might affect crop growth are cadmium and chromium. Experiments with cadmium salts and with sewage sludge spiked with cadmium have shown that it is much more phytotoxic than zinc, but the concentrations normally found in sewage sludge are relatively low so that in practice it is unlikely to cause growth problems in crops. Chromium, especially when used in Cr^{6+} form, has been found to produce toxic effects on crops in pot experiments, but there is no evidence that it is harmful to crop growth when added to the soil in sewage sludge, where it occurs in Cr^{3+} form and is only slightly soluble in extractants normally used for measuring 'availability' to crops. This is in agreement with the work on tannery sludges containing very high chromium concentrations reported above (Mazur and Koc, 1976*a*,*b*).

Two experiments were started in England in 1968 to compare the effects on a range of vegetable crops of four sewage sludges in which the predominant metals were zinc, copper, nickel and chromium, and the results have now been reported (Marks *et al.*, 1980). Here, no toxic effects on crops followed the use of the sludge with high chromium content despite the very heavy maximum application rate used (125 t/ha with 8600 ppm chromium). The results obtained for the other three metals are difficult to interpret because the sludges with high copper and nickel contents also contained large amounts of zinc, and because the heavy application rates used for the highly contaminated sludges often caused complete crop failure. However, by using multiple regression techniques on the data, the relative toxicities for the metals zinc, copper and nickel were calculated as given in Table 5.14. Nickel was confirmed as the most toxic of the three for

TABLE 5.14

RELATIVE TOXICITIES TO CROPS OF TRACE METALS ADDED TO THE SOIL IN SEWAGE SLUDGE. (AFTER MARKS *et al.*, 1980)

Site	Crop	Zinc	Copper	Nickel
1	Red beet	1	0·75	1·75
	Celery	1	0	4·88
2	Red beet	1	1·31	2·16
	Lettuce	1	0·87	1·77

all the crops tested but the effect of copper was more variable, celery being very tolerant and the other crops, red beet and lettuce, showing effects not very different from that due to zinc. Bingham *et al.* (1979), using spiked sludges, compared the toxicities of zinc, copper, nickel and cadmium to spring wheat grown in pots. Under acid conditions toxic effects on growth were in the ratio 1:1·4:2:4 for the four elements, but after heavy liming only copper and cadmium produced any reduction in yield. Crops evidently vary greatly in their sensitivity to trace metals in the soil which makes the calculation of application limits very difficult except on the basis that they should allow even the most vulnerable crop to be grown safely.

Trace metals in sewage sludge which could affect agriculture by rendering crops unfit for human consumption include cadmium, lead and mercury, all cumulative and toxic in man. Other elements, such as arsenic, which form poisonous compounds, may also need to be considered occasionally, although concentrations in sewage sludge are generally low, and uptake by crops is also limited. At the present time, cadmium is perhaps the element causing most concern, because levels of intake in the average diet are higher in relation to those levels thought to be safe over a long period, as laid down by the World Health Organisation, than for any other element. Andersson (1976) has argued that additions of cadmium to the soil in fertilizers and manures should be kept down to 4 g/ha/year which he calculated was the maximum amount likely to be added in farmyard manure and a rate which would result in no increase in soil cadmium content. Cadmium is well known to be freely taken up and translocated by plants, so that increases in soil cadmium result in increased crop contents. Only small amounts appear to accumulate in grain. Andersson and Nilsson (1976) showed that after a total application of about 120 t/ha sewage sludge dry matter over 17 years, which increased soil cadmium soluble in 2 M nitric acid from 0·24 to 0·72 ppm, grain cadmium was raised from 0·036 to 0·068 ppm. The sludge used had a cadmium content of 10–13 ppm dry matter. In another long-term experiment with maize, Hinesly *et al.* (1979) applied 232 t/ha sludge dry matter containing 58 kg cadmium over 5 years. The cadmium content of the grain rose to a maximum of 1·37 ppm in the third year of the experiment but rapidly declined, so that 3 years after sludging was discontinued concentrations were similar to those for grain from the unsludged plots (0·10–0·15 ppm). They suggested that the availability of cadmium added to the soil decreased with time. Baerug and Martinsen (1977) have reviewed the data for potatoes and concluded that cadmium in the tubers was unlikely to be a hazard to consumers.

Vegetable crops, especially those where the leaves are eaten, may have higher cadmium concentrations than those found in grain or tubers. Wood *et al.* (1978) presented data for an area west of London where sewage sludge with over 100 ppm cadmium in the dry matter had been applied to market garden land for many years. Soil cadmium contents up to 24 ppm were found with concentrations in lettuce dry matter up to 10 ppm. John and van Laerhoven (1976) have shown that different cultivars of lettuce grown in the same medium can contain widely varying concentrations of cadmium, which suggests that the breeding of strains with low metal uptake might be a way of allowing the safe cropping of polluted soils in future. There is general agreement that additions of cadmium to the soil in sewage sludge must be kept under control although views differ, as shown below, as to the limits which should be applied.

Sewage sludge contains appreciable amounts of lead; in different investigations the median concentration has varied from 180 to 1450 ppm in the dry matter. Lead applied to the land in sewage sludge appears to be taken up by crops to a very limited extent, and most experiments have failed to demonstrate any significant increase in the lead content of food crops. In relation to animal nutrition, ingestion of contaminated soil may be a hazard if the lead concentration of the surface layers becomes too high, while direct ingestion of sewage sludge adhering to grass could also be harmful to livestock if lead is present in high concentrations. In this connection the work of Chaney *et al.* (1977) is significant. They showed that tall fescue could remain physically contaminated for 80 days even though 41 mm of rain had fallen and suggested that the drop in metal content of the grass with time was due to dilution by additional growth rather than to the sludge being washed off by the rain. In general, however, additions of lead to the soil in sewage sludge are not likely to present a hazard except, perhaps, in areas where the soils are already contaminated with lead from other sources.

Mercury concentrations in sewage sludge rarely exceed 50 ppm in the dry matter and in the studies summarized by Berrow and Burridge (1980) median values were all in the range 3–7 ppm. Some mercury applied in sewage sludge is likely to be lost to the atmosphere and the element is only taken up by plants to a limited extent. Andersson and Nilsson (1976) in the long-term experiment described above found that soil mercury extractable with 2 M nitric acid had risen from 0·05 to 0·75 ppm after 17 years' application of sludge containing about 9 ppm mercury in the dry matter, but the concentration in grain was still below the limit of detection (0·005 ppm). Despite its toxic nature, mercury in sludge does not appear to present problems to agriculture.

Molybdenum in sewage sludge can cause problems. High levels in the soil can result in hypocuprosis in cattle and this has been recognized in one area in England where there had been continued use of a sludge with a high molybdenum concentration on grassland. The average molybdenum content of sewage sludge dry matter is about 5 ppm but concentrations up to 30–40 ppm have been recorded. Most sludges contain much more copper than molybdenum and it might be expected that levels of copper in herbage would increase at least as much as those of molybdenum when sludge was applied. In some cases, especially on acid soils, this may be the case, but under alkaline conditions molybdenum uptake is enhanced while that of copper, which is not readily translocated from roots to shoots, is depressed. Care needs to be taken when using sludges with 20–40 ppm molybdenum especially when they have been stabilized with lime, or on soils with pH values of 7·0 or above.

In recent years, increasing concern about additions of trace metals in sewage sludge to the soil has led to the formulation of guidelines which, if followed, should prevent adverse effects on agriculture. The first of these (Chumbley, 1971) for England and Wales only dealt with elements thought likely to affect crop growth, but they have now been extended by the Standing Committee on the Disposal of Sewage Sludge (Dept. Environment, 1981) to cover a much wider range of trace elements. In order to simplify the calculation of amounts of sludge which could safely be used, Chumbley introduced the concept of 'zinc equivalent' based on analytical data. This was calculated from the formula: zinc equivalent = zinc + 2 × copper + 8 × nickel; the figures in all cases being concentrations in ppm of sludge dry matter. It was assumed from current knowledge that 280 kg/ha of 'zinc equivalent' could safely be applied over a long period, taken for purposes of calculation to be 30 years, provided soil pH was maintained at 6·5 or above. This concept has been retained in the latest British guidelines, although as shown by Marks *et al.* (1980) and Bingham *et al.* (1979) the relative toxicities of the three elements vary with the crop and with soil pH value. Also, on theoretical grounds, the effects of the three metals are unlikely to be strictly additive as implied by the use of 'zinc equivalent'.

Guidelines have also been produced in other countries and some of these are summarized in Table 5.15. There are evidently major differences of opinion as to what can be considered as 'safe' levels of application, despite the large volume of work published during the last decade. The variations not only reflect differences in attitude from one country to another, but also the real differences between soils, crops and climate which can affect the results of research. Much of the work done on trace metals has tested additions of metal salts either directly or added to sewage sludge, and because of the numbers of treatments involved, crops have often been grown in pots. Recently De Vries and Tiller (1979) have shown that results from pot experiments can be quite different from those obtained with the same treatments in the field. Comparisons of metals applied as simple salts and in sewage sludge have also been shown to give widely varying results. Calculation of safe additions of trace metals to the soil assumes that the total amounts applied may affect crops either immediately or as they become available in future. Few long-term investigations have been completed but here too there are differences of opinion from the results obtained. Hinesly *et al.* (1979) showed a rapid decline in grain cadmium and zinc contents after sludge application stopped, while Berrow and Burridge (1980) argued

TABLE 5.15

COMPARISON OF GUIDELINES FOR METAL APPLICATION TO AGRICULTURAL LAND IN SEWAGE SLUDGE (kg/ha)

Trace element	Great Britain	Ontario	US	Sweden	Holland
Zinc	560	363	500	300	120
Copper	280	168	250	90	30
Nickel	70	36	100	15	3
Chromium	1 000	220	—	30	—
Cadmium	5	1·6	10	0·75	0·6
Lead	1 000	94	1 000	30	—
Mercury	2	0·9	—	0·75	0·6
Molybdenum	4	3·8	—	—	—

Great Britain (Dept. Environment, 1981): Zinc, copper and nickel contents totalled by 'zinc equivalent' formula. Soil pH 6·5 or above.
Ontario (Webber, 1979): Soil pH 6·0 or above.
US (US Fed. Register, 1977, 1978): Soil pH 6·5 or above, rates for soils with cation-exchange capacity 5–15 meq/100 g, halved where <5 meq/100 g, doubled where >15 meq/100 g.
Sweden (Tullander, 1975): Based on annual rate × 30.
Holland (De Haan, 1974): Based on annual rate × 30.
Rates are for arable land and may be modified for grassland, etc.

from measurements of uptake by grass 8 years after sludge was applied and from the extractable metal contents of the soil that there was little evidence of any reduction in the toxicity of zinc, copper and nickel with time.

5. CONCLUSIONS

Many problems remain to be solved in relation to trace metals and agriculture, since for elements like cadmium and lead, which have no known essential biological function, any addition to the soil is potentially harmful and the aim must be to keep such additions down to the minimum possible. However, a balance has to be struck between the needs of industry to use such trace metals and the difficulties and cost of preventing them reaching the land by one or more of the routes already described. In the case of other elements some addition to the soil may, where there are deficiencies, be beneficial but when the amounts added become large similar problems are posed. Where trace metals are added in agricultural practices alternatives need to be sought and this is being done; the replacement of copper fungicides by others with no metals in them is an example of this. Further long-term studies on soils and crops are needed, together with continuous vigilance to ensure that pollution is kept to the minimum at all points. A fresh approach to guidelines or codes of practice based on the levels found in soil rather than on additions of particular pollutants may also be desirable in that it would look at all forms of pollution which might affect crop growth or composition. It might even be easier to agree on maximum safe soil concentrations than on permissible addition of trace metals. Much obviously remains to be done in this area of research.

REFERENCES

Anderson, R. J., B. E. Davies, J. H. Nunn and P. M. C. James (1979). The dental health of children from five villages in Somerset with reference to environmental cadmium and lead. *Brit. Dent. J.* **147**: 159–61.

Andersson, A. (1976). On the influence of manure and fertilisers on the distribution and amounts of plant-available cadmium in soils. *Swedish J. Agric. Res.* **6**: 27–36.

Andersson, A. and K. O. Nilsson (1976). Influence on the levels of heavy metals in soil and plant from sewage sludge used as fertiliser. *Swedish J. Agric. Res.* **6**: 151–9.

Baerug, R. and J. H. Martinsen (1977). The influence of sewage sludge on the content of heavy metals in potatoes and on tuber yield. *Plant Soil.* **47**: 407–18.

Batey, T., C. Berryman and C. Line (1972). The disposal of copper-enriched pig manure on grassland. *J. Br. Grassl. Soc.* **27**: 139–43.

Berrow, M. L. and J. C. Burridge (1980). Trace element levels in soils: Effects of sewage sludge. In: *Inorganic Pollution and Agriculture*, MAFF Ref. Book 326, HMSO, London, pp. 159–83.

Berrow, M. L. and J. Webber (1972). Trace elements in sewage sludges. *J. Sci. Fd. Agric.* **23**: 93–100.

Berryman, C. (1970). The problem of disposal of farm wastes with particular reference to maintaining soil fertility. *Symposium Farm Wastes*, University of Newcastle-upon-Tyne, UK, Jan. 1970, pp. 19–23.

Bingham, F. T., A. L. Page, G. A. Mitchell and G. E. Strong (1979). Effects of liming on acid soil amended with sewage sludge enriched with cadmium, copper, nickel and zinc on yield and cadmium content of wheat grain. *J. Environ. Qual.* **8**: 202–7.

Bishop, R. F. and D. Chisholm (1962). Arsenic accumulation in Annapolis Valley orchard soils. *Can. J. Soil Sci.* **42**: 77–80.

Burkitt, A., P. Lester and G. Nickless (1972). Distribution of heavy metals in the vicinity of an industrial complex. *Nature, Lond.* **238**: 327–8.

Chaney, R. L., S. B. Hornick and P. W. Simon (1977). Heavy metal relationships during land utilisation of sewage sludge in the North East. In: *Land as a Waste Management Alternative* (R. C. Loehr (ed)), Ann Arbor Science Publishers, Ann Arbor, Michigan, pp. 285–314.

Chisholm, D. (1972). Lead, arsenic and copper content of crops grown on lead arsenate treated and untreated soils. *Can. J. Plant Sci.* **52**: 583–8.

Chumbley, G. C. (1971). Maximum permissible levels of metals in sewage applied to agricultural land. ADAS Adv. Paper No. 10, MAFF, London.

Davies, B. E. (1971). Trace metal content of soils affected by base metal mining in the west of England. *Oikos.* **22**: 366–72.

Davies, B. E. (1976). Mercury content of soils in western Britain with special reference to contamination from base metal mining. *Geoderma.* **16**: 183–92.

Davies, B. E. (1980). Base metal mining and heavy metal contamination of agricultural land in England and Wales. In: *Inorganic Pollution and Agriculture*, MAFF Ref. Book 326, HMSO, London, pp. 142–56.

Davies, B. E. and P. A. Holmes (1972). Lead contamination of roadside soil and grass in Birmingham, England in relation to naturally occurring levels. *J. Agric. Sci., Camb.* **79**: 479–84.

De Haan, S. (1974). Die chemische Zusammensetzung von Gewächsen auf mit Klärschlamm behandelten Boden. *Landwirtsch. Forsch., Sonderh.* **31**: 220–33.

De Vries, M. P. C. and K. G. Tiller (1978). Sewage sludge as a soil amendment with special reference to Cd, Cu, Mn, Ni, Pb and Zn—comparison of results from experiments conducted inside and outside a glasshouse. *Environ. Pollut.* **16**: 231–40.

Delas, J. (1963). La toxicité du cuivre accumulé dans les sols. *Agrochimica.* **7**: 258–87.

Dept. Environment (1976). Environmental mercury and man. Department of the Environment, Pollution Paper No. 10, HMSO, London.

Dept. Environment (1978). Sewage sludge disposal data and reviews of disposal to sea. Standing Tech. Committee Report. No. 8, Department of the Environment, HMSO, London.

Dept. Environment (1981). Report of the Sub-Committee on the Disposal of Sewage Sludge to Land. Standing Tech. Committee Report. Department of the Environment, HMSO, London (in press).

Fleming, G. A. (1962). Selenium in Irish soils and plants. *Soil Sci.* **94**: 28–35.

Gracey, H. I., T. A. Stewart, J. D. Woodside and R. H. Thompson (1976). The effect of disposing of high rates of copper-rich pig slurry on grassland on the health of grazing sheep. *J. Agric. Sci., Camb.* **87**: 617–23.

Gray, K. R. and A. J. Biddlestone (1980). Agricultural use of composted town refuse. In: *Inorganic Pollution and Agriculture*, MAFF Ref. Book 326, HMSO, London, pp. 279–305.

Griffiths, J. R. and G. A. Wadsworth (1980). Heavy metal pollution on farms near an industrial complex. In: *Inorganic Pollution and Agriculture*, MAFF Ref. Book 326, HMSO, London, pp. 70–6.

Hinesly, T. D., E. L. Ziegler and G. L. Barrett (1979). Residual effects of irrigating corn with digested sewage sludge. *J. Environ. Qual.* **8**: 35–8.

John, M. K. and C. J. van Laerhoven (1976). Differential effects of cadmium on lettuce varieties. *Environ. Poll.* **10**: 163–73.

Kimura, Y. and V. L. Miller (1964). The degradation of organo–mercury fungicides in soil. *J. Agric. Fd. Chem.* **12**: 253–7.

Klein, L. A., M. Lang, N. Nash and S. L. Kirschner (1974). Sources of metals in New York City wastewater. *J. Wat. Poll. Cont. Fed.* **46**: 2653–62.

Kneale, W. A. and P. Smith (1977). The effect of applying pig slurry containing high levels of copper to sheep pastures. *Expl. Husb.* **32**: 1–7.

Koc, J., L. Krefft and T. Mazur (1976). Investigations into the fertilising value of tannery sludges. I. Chemico-physical characteristics of sludges. *Roczniki Gleboznawcze.* **27**: 103–22.

Kornegay, E. T., J. D. Hedges, D. C. Martens and C. Y. Kramer (1976). Effect of soil and plant mineral levels following application of manures of different copper contents. *Plant Soil.* **45**: 151–62.

Lagerwerff, J. V. (1971). Uptake of cadmium, lead and zinc by radish from soil and air. *Soil Sci.* **111**: 129–33.

Le Riche, H. H. (1959). Molybdenum in the Lower Lias of England in relation to the incidence of teart. *J. Soil Sci.* **10**: 133–7.

Lee, K. W. and D. R. Keeney (1975). Cadmium and zinc additions to Wisconsin soils by commercial fertilisers and wastewater sludge applications. *Water Air Soil Poll.* **5**: 109–12.

Marks, M. J., J. H. Williams, L. V. Vaidyanathan and C. J. Chumbley (1980). Field experiments testing the effects of metal-contaminated sewage sludges on some vegetable crops. In: *Inorganic Pollution and Agriculture*, MAFF Ref. Book 326, HMSO, London, pp. 235–46.

Martin, M. H., P. J. Coughtrey, S. W. Shales and P. Little (1980). Aspects of airborne cadmium contamination of soils and natural vegetation. *In: Inorganic Pollution and Agriculture*, MAFF Ref. Book 326, HMSO, London, pp. 56–69.

Mazur, T. and J. Koc (1976*a*). Investigations into the fertilising value of tannery sludge. III. Effect of tannery sludge fertilising on the chemical composition of crops. *Roczniki Gleboznawcze.* **27**: 123–35.

Mazur, T. and J. Koc (1976*b*). Investigations into the fertilising value of tannery sludge. IV. Effect of fertilising with tannery sludges on changes in chemical properties of soil. *Roczniki Gleboznawcze.* **27**: 137–46.

Mitchell, R. L. (1974). Trace element problems in Scottish soils. *Neth. J. Agric. Sci.* **22**: 295–304.

Mitchell, R. L. and J. W. S. Reith (1966). Lead content of pasture herbage. *J. Sci. Fd. Agric.* **17**: 437–40.

Morrison, J. L. (1969). Distribution of arsenic from poultry litter in broiler chickens, soil and crops. *J. Agric. Food Chem.* **17**: 1288–93.

Page, A. L. and T. J. Ganje (1970). Accumulations of lead in soils for regions of high and low motor vehicle density. *Environ. Sci. Technol.* **4**: 140–2.

Patterson, J. B. E. (1971). Metal toxicities arising from industry. In: *Trace Elements in Soils and Crops*, MAFF Tech. Bull. 21, HMSO, London, pp. 193–207.

Porter, E. K. and P. J. Peterson (1975). Arsenic accumulation by plants on mine waste (United Kingdom). *Sci. Total Environ.* **4**: 365–71.

Purves, D. (1972). Consequences of trace element contamination of soils. *Environ. Pollut.* **3**: 17–24.

Purves, D. and E. J. Mackenzie (1973). Effects of applications of municipal compost on uptake of copper, zinc and boron by garden vegetables. *Plant Soil.* **39**: 361–71.

Purves, D. and E. J. Mackenzie (1974). Phytotoxicity due to boron in municipal compost. *Plant Soil.* **40**: 231–5.

Rohde, G. (1962). The effects of trace elements on the exhaustion of sewage-irrigated land. *J. Inst. Sewage Purif.* **61**: 581–5.

Saha, J. G., Y. W. Lee, R. D. Tinline, S. H. F. Chinn and H. M. Austenson (1970). Mercury residues in cereal grains from seeds or soil treated with organo–mercury compounds. *Can. J. Plant Sci.* **50**: 597–9.

Shewry, P. R. and P. J. Peterson (1976). Distribution of chromium and nickel in soils from Serpentine and other sites. *J. Ecol.* **64**: 195–212.

Stenström, T. and M. Vahter (1974). Cadmium and lead in Swedish commercial fertilisers. *Ambio.* **3**: 91–2.

Stewart, W. L. and R. Allcroft (1956). Lameness and poor thriving in lambs on farms in old lead mining areas in the Pennines. I. Field investigations. *Vet. Rec.* **68**: 723–8.

Thomas, R. (1980). Arsenic pollution arising from mining activities in south-west England. In: *Inorganic Pollution and Agriculture*, MAFF Ref. Book 326, HMSO, London, pp. 126–41.

Thompson, I., I. Thornton and J. S. Webb (1972). Molybdenum in black shales and the incidence of bovine hypocuprosis. *J. Sci. Fd. Agric.* **23**: 878–91.

Thornton, I. (1974). Biogeochemical and soil ingestion studies in relation to the trace element nutrition of livestock. In: *Trace Element Metabolism in Animals—2* (W. G. Hockstra, J. W. Suttie, H. E. Ganther and W. Mertz (eds)), University Park Press, Baltimore, pp. 451–4.

Thornton, I. (1980). Geochemical aspects of heavy metal pollution and agriculture in England and Wales. In: *Inorganic Pollution and Agriculture*, MAFF Ref. Book 326, HMSO, London, pp. 105–25.

Tullander, V. (1975). Final disposal of municipal sludge in Sweden. *J. Wat. Poll. Cont. Fed.* **47**: 688–95.

Unwin, R. J. (1980). Copper in pig slurry: Some effects and consequences of spreading on grassland. In: *Inorganic Pollution and Agriculture*, MAFF Ref. Book 326, HMSO, London, pp. 306–19.

US Fed. Register (1977). Municipal sludge management: Environmental factors. *Federal Register.* **42**: 57420–7.

US Fed. Register (1978). Solid waste disposal facilities. Proposed classification criteria. *Federal Register.* **43**: 4942–55.

Webber, M. D. (1979). Canadian research on land utilisation of sewage sludge. European Symposium on the Characterisation, Treatment and Use of Sewage Sludge, Cadarache, France, Feb. 1979.

Williams, C. H. and D. J. David (1973). The effect of superphosphate on the cadmium content of soil and plants. *Aust. J. Soil Res.* **11**: 43–56.

Williams, R. O. (1975). A survey of the heavy metal and inorganic content of sewage sludges. *Wat. Pollut. Control.* **74**: 607–8.

Wood, L. B., R. P. King and P. E. E. Norris (1978). Some investigations into sludge-amended soils and associated crops and the implications for trade effluent control. Symposium on Utilisation of Sewage Sludge on Land, Oxford, England, April 1978.

CHAPTER 6

Reclamation of Metalliferous Mine Wastes

A. WILLIAMSON and M. S. JOHNSON
Department of Botany, University of Liverpool, UK

1. INTRODUCTION

During the last decade there has been a dramatic worldwide increase in the quantities of mineral wastes produced by the metalliferous mining industry. This has resulted from a combination of social, economic and technological factors, in particular the ever-increasing requirement for metalliferous minerals and the ability of modern mining and processing methods to economically exploit ore deposits containing very low concentrations of valuable metals. Extraction of low-grade orebodies including as little as 0·2 % w/w of recoverable metal inevitably generates large amounts of waste which are commonly discarded in land-based impoundments or tips, and may cover several square kilometres from a single mining venture.

Mineral waste disposal presents a broad spectrum of potential environmental problems ranging from pollution of terrestrial and aquatic ecosystems to conflicts associated with landscape quality. Responding to past experience, and in an attempt to avoid problems in the future, several countries have evolved legislation on environmental protection which requires a detailed submission of intended land reclamation procedures at the mine planning stage. This policy has been extended by some authorities who require, before planning permission is granted, a substantial bond which is returnable on completion of a mutually agreed pollution control and land reclamation plan.

The establishment and maintenance of vegetation is usually an integral part of reclamation programmes for metalliferous wastes. However, successful revegetation requires a thorough understanding of the physico-chemical and biological constraints upon plant growth, and the means by which they may be overcome, in order to minimize the long term hazards to environmental quality posed by derelict mineral workings.

2. ORIGIN AND DISPOSAL OF WASTES

Waste materials produced at individual sites vary with the nature of the mining programme, and include waste rock, overburden, mill tailings and liquors. To a considerable extent, underground operations are able to exploit selected seams containing the required grade of ore, thus avoiding less suitable material. The crude ore is finely ground before extraction of the valuable metallic minerals, and the milled waste, known as tailings, is then discarded in the form of an aqueous slurry. This contrasts with open-pit workings where total excavation of the deposit occurs. Material with a metal content above the acceptable cut-off grade is ground and processed whilst that below the current threshold is stockpiled or permanently discarded. Small quantities of the latter are used in the construction of

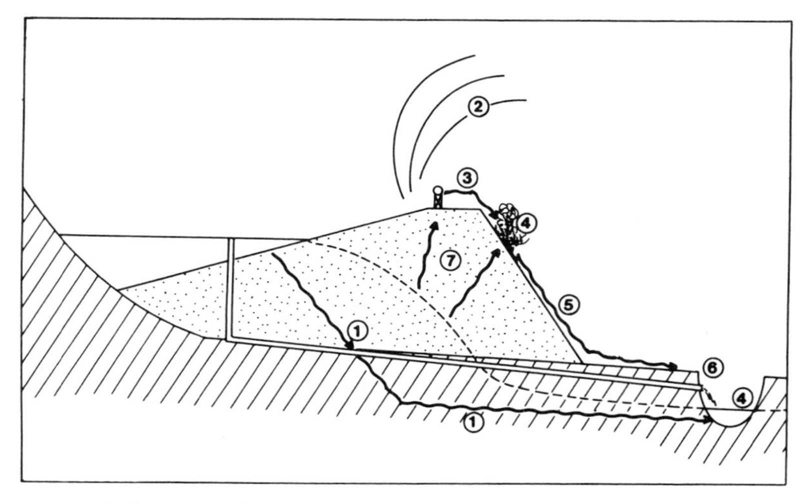

FIG. 6.1 Pollution pathways from tailings. (Adapted from Andrews, 1975.) 1 Water pollution through seepage, 2 dust blow, 3 operational failure, 4 biological accumulation of toxic metals, 5 erosion or slumping, 6 decant discharge, 7 capillary transport of metals.

amenity banks and tailings dam walls, but the majority is dumped adjacent to the pit.

Various methods of tailings disposal have been used throughout the world, including discharge to lakes, rivers, estuaries and the open sea. Generally, the method adopted is determined by location and environmental considerations. For example, river disposal is employed at several mines in the Philippines (Down and Stocks, 1977*b*), and at one of the world's major copper mines, Bougainville, Papua New Guinea, where 80 000–90 000 t of tailings/day are discarded in this manner (Hartley, 1976). Environmental damage is inevitable in such cases, but in an area subject to frequent earthquakes there is little opportunity for safe disposal on land. The method of tailings disposal most commonly used worldwide is land impoundment. Containment dams have received universal acceptance by the mining industry as the principal means for storage of mill wastes, since they provide a relatively inexpensive way of treating and disposing of tailings and liquors and provide for a reprocessing option in the future (Bell, 1974). Nevertheless, serious environmental problems may derive from tailings dams (Fig. 6.1), requiring implementation of pollution control measures including temporary surface stabilization and permanent rehabilitation on completion of disposal operations.

3. STABILIZATION METHODS

Successful stabilization of waste dumps and tailings dams begins at the design stage. Reduction in the length of slopes, terracing or staircase grading, help to reduce susceptibility to erosion, and facilitate later establishment of vegetation (Kay, 1977).

The US Bureau of Mines has pioneered chemical stabilization techniques and identified various compounds of use on certain toxic mineral wastes, particularly when used in conjunction with vegetation establishment (Dean and Havens, 1973). The most successful materials include petroleum-based products, resin emulsions, lignosulphonates, bitumen-based materials and organic polymers (Shirts and Bilbrey, 1976). Alternative systems based on physical methods of stabilization include temporary flooding of beached areas of active dam surfaces or covering of abandoned sites with innocuous native rock or soil (Dean and Havens, 1973).

However, after extensive worldwide experimentation with physical, chemical and biological methods, the favoured stabilization method is by the use of vegetation. This has the potential for aesthetic improvements, long-term surface stability, the development of ecological and wildlife interest and a low maintenance commitment. Moreover it is compatible with reprocessing of the wastes if this should eventually prove to be economically viable.

4. RECLAMATION OBJECTIVES

Development of a reclamation plan for a mine requires a thorough understanding of the local climate, topography, geology, soils, hydrology, flora and fauna as well as knowledge of the local mining and planning laws. After-uses that have been successfully applied to mine sites include: (1) development of reserves and areas of ecological interest following reinstatement of the native vegetation; (2) agricultural development; (3) forestry; and (4) amenity and recreational facilities such as camping grounds, fishing lakes and golf courses. However, increasing appreciation of the potential hazards of redeveloping contaminated areas for utilization by the public seriously questions the wisdom of using disused mine sites for certain of these objectives. In extreme climates revegetation using well-adapted native species may be advantageous. Native species provide an aesthetically attractive vegetation cover, create potentially valuable wildlife habitats and in certain areas of the world may be required to restore ancient rights of the people, e.g. in areas of aboriginal cultures (Middleton, 1979).

In the majority of cases, physical and chemical limitations, particularly toxicity of the wastes, will preclude agricultural management. Continuous grazing by livestock is often inadvisable on reclaimed lands due to the toxicity hazards of heavy metals in the herbage (Down, 1974). If grazing is permitted under controlled conditions, such as short-term rotational systems, regular sampling and analysis of vegetation is essential to monitor the biologically significant trace metals (e.g. cadmium, lead and mercury) which may be present in foliar tissues at elevated concentrations (Johnson and Bradshaw, 1977). Less-toxic wastes can be satisfactorily restored to agriculture or forestry, if adequate care is taken to provide a physically suitable substrate. Thus the disused opencast tin mines of the Jos Plateau in Nigeria have been levelled and the innocuous but nutrient-poor subsoil induced to yield successful plantations of eucalyptus to meet local timber requirements (Wimbush, 1963). In Canada, the International Nickel Company has marketed several thousand bales of hay, made from grass established on nickel mine tailings (Peters, 1970).

5. CONSTRAINTS TO PLANT GROWTH

The constraints to growth of a particular mine waste should be investigated, as far as possible, before a reclamation plan is formulated. The general properties of the different types of waste encountered in mining are outlined in Tables 6.1 and 6.2. Overburden, which may contain the topsoil fraction, can deteriorate physically through incorrect handling and storage, but in general, low fertility and an inherent poor physical structure are the main plant growth problems. Waste rock is usually physically stable but

TABLE 6.1
MAJOR PLANT GROWTH LIMITATIONS OF METAL MINE WASTES

Waste material	Texture and structure	Stability	Water supply	Nutrient status	pH	Toxic metals	Salinity
Overburden	0–00	0–0	0–0	0–00	0–00	0	0
Waste rock	000	0–0	000	000	0–000	0–000	0–00
Tailings	00	0–000	0	000	0–000	0–000	0–000

Severity of problem: 0, insignificant; 0, slight; 00, moderate; 000, severe.

nevertheless has a high capacity for contamination of surrounding ecosystems through weathering reactions leading to percolation and dispersion of toxic metals. Without amelioration, waste rock does not provide a suitable matrix for plant growth as it lacks essential plant nutrients and is subject to moisture stress due to the absence of water-retaining particles of fine sand, silt and clay dimensions (Table 6.2).

Tailings are composed of small and uniform particles, mainly in the sand and silt categories. This limited size range and absence of the principal agents that control aggregation, namely organic matter, clay minerals and microorganisms, renders the material structureless, unstable and subject to erosion (Burton *et al.*, 1978). Porosity, aeration, water infiltration and percolation are frequently unsatisfactory for plant growth. The bare, unprotected surfaces are exposed to extremes of temperature which accentuates their susceptibility to erosion. Moreover, surface crusting, compaction, and high bulk density are inhibitory to root penetration and growth.

In terms of chemical composition, mine tailings often contain phytotoxic levels of heavy metal ions, residual quantities of process reagents, chemical

TABLE 6.2

ANALYTICAL DATA FROM CANADIAN METAL MINE WASTES. (ADAPTED FROM MURRAY, 1977)

Parameter	Overburden[a]		Waste rock[b]		Tailings[c]	
	Mean	Range	Mean	Range	Mean	Range
Electrical conductivity (mmho/cm)	0·7	0·1–3·1	0·9	0·3–3·5	2·0	0·1–22·4
Organic matter (%)	4·4	0·3–6·6	2·5	0·1–19·5	2·0	0·02–25
Particle size (% <2 mm)	38	8–95	24	10–78	95	20–100
Available water storage capacity (%)	12	5–19	9	3–17	16	0–35
Cation-exchange capacity (meq/100 g)	16	1–118	11·4	0·3–32·4	2·63	0·19–46·5
Bulk density (g/cm^3)	1·42	1·10–1·99	2·04	1·19–3·02	1·5	0·2–3·1
Available P (μg/g)	7·3	0·2–23	4·4	0–33·4	10	0·1–400
Available K (μg/g)	70	15–595	85	4–193	63	1–564
Total N (%)	0·17	0·01–0·87	0·01	0–0·12	0·013	0·001–0·166
Total S (%)	1·40	0·34–7·96	0·04	0–0·15	4·02	0·01–38·87

[a] Data from 27 or more observations from 8 mine sites.
[b] Data from 39 or more observations from 11 mine sites.
[c] Data from 84 or more observations from 43 mine sites.

complexes formed during processing and a high level of dissolved solids (Bell, 1975). The exact chemical composition is complex and variable, depending not only on the nature of the original ore and milling reagents added, but also on climatic conditions and on weathering reactions that occur following disposal.

5.1. Acidity

The major chemical problem in many modern metalliferous mine spoils is the generation of acidity induced by the weathering of sulphides, particularly iron pyrites (FeS_2). In general, substrates with a pH 5·5–7·0 are required for satisfactory plant growth, with an optimum of pH 6·5. Values of less than 5·0 inhibit normal development of most plant species. Certain mining operations bring to the surface pyritic materials which on exposure to air and water undergo a series of oxidation and hydrolysis reactions assisted by ferrous ion oxidizing bacteria (e.g. *Thiobacillus ferro-oxidans*). The degradation and dissolution of pyrite may be summarized as:

$$4FeS_2 + 15O_2 + 14H_2O \rightarrow 4Fe(OH)_3 + 8H_2SO_4$$

Secondary reactions between the products of weathering and associated heavy metal sulphides in the wastes, produce soluble and mobile metal sulphates by a series of reactions which can be abbreviated to:

$$(M)S + 2Fe_2(SO_4)_3 + 2H_2O + 3O_2 \rightarrow (M)SO_4 + 4FeSO_4 + 2H_2SO_4$$

Moreover, the phytotoxic accessory metals, manganese and aluminium, are brought into solution during this process. The pyrite content of wastes is therefore vitally important in determining the difficulties liable to be encountered in revegetation; wastes with a low pyrite content presenting fewer problems. However, a high level of pyrite does not automatically produce a highly acidic waste. The factors liable to influence acidification are the carbonate to pyrite ratio of the material, since native carbonates act as neutralizers of generated acidity, and the reactivity of the pyrite itself, which is related at least in part to the crystal form (Costigan, 1979). In climates where rainfall exceeds evaporation, leaching transports soluble metal ions and acidity through the tailings profile. This principle has been utilized by installing artificial irrigation systems in some South African gold tailings reclamation schemes (James, 1966). In a study of weathered tailings, Boorman and Watson (1976) found that oxidation of pyrites occurred down to 50 cm beneath the surface and that the acidity in this zone facilitated leaching of the soluble metal ions down to 50 cm depth where a hard pan of iron oxides and heavy metals formed (Fig. 6.2).

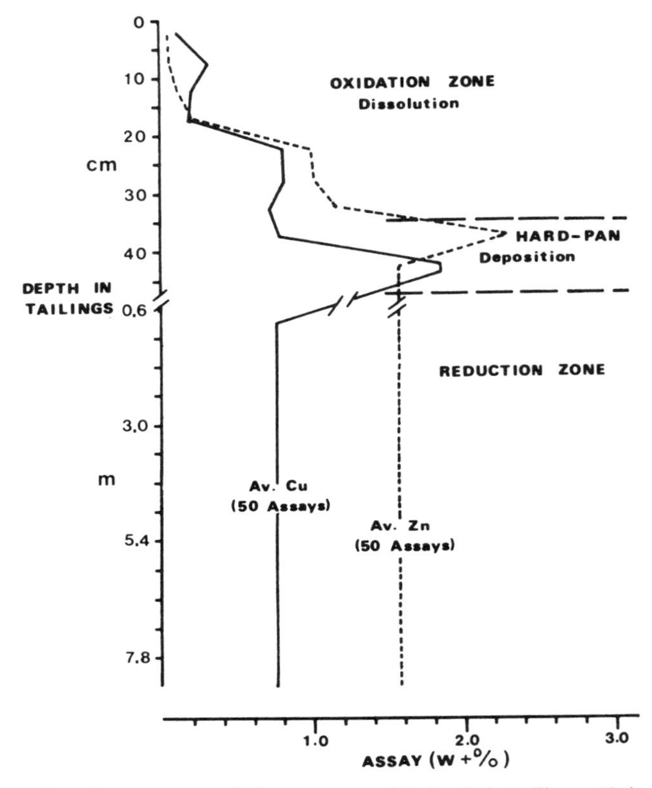

FIG. 6.2 Changing copper and zinc content with depth in tailings. (Adapted from Boorman and Watson, 1976.)

5.2. Toxic Metals

In general, metal concentrations above 0·1 %w/w prove phytotoxic (Bradshaw and Chadwick, 1980), though this depends on the metallic species, release characteristics and the nature of the accessory minerals. Modern extraction processes produce wastes with metal levels of this order, whereas spoils from older operations (Table 6.3), with less-efficient recovery methods, frequently contain between 1 % and 3 %w/w of the metals and in some cases even higher concentrations (Johnson and Bradshaw, 1977).

Associated metals including arsenic, cadmium, antimony and silver may also be present at concentrations above the normal range for agricultural soils (Swaine, 1955). Their toxicity and ecological importance vary but cadmium, for example, an element commonly associated with zinc in sphalerite, may behave in a manner analogous to persistent pesticides in distribution, accumulation and concentration in food chains (Goodman *et al.*, 1973). Tailings may also contain residual flotation reagents from mill processing. Cyanides, silico-fluorides and ammoniacal liquors are of paramount importance in terms of environmental damage (Bell, 1975). In most cases, concentrations of flotation reagents must exceed 25 mg/litre, and in some instances 200 mg/litre in drinking water, before toxic effects are exhibited in man (Duncan, 1972).

5.3. Salinity

High salinity is a common feature of modern tailings and conductivities above 16 mmho/cm, the level above which no crops survive, have been reported in recently discarded material (Nielson and Peterson, 1972). This results from interactions between the products of pyrite weathering and native carbonates (e.g. $MgCO_3$), from concentration of naturally occurring salts by recycling of water in mill processes, and also from chemical additives used to adjust effluent pH prior to discharge. At Mount Isa Mines in Australia, where evaporation greatly exceeds rainfall throughout the year, high surface salt concentrations cause major problems for plant establishment (Farrell, 1977).

5.4. Nutrient Status

A universal constraint to plant growth on waste rock, overburden and tailings dams is the low concentration of essential plant nutrients. This reflects severe deficiencies or an absence of organic matter and clay minerals which provide the nutrient store and cation-exchange capacity of normal soils. Nitrogen, phosphorus and, to a lesser extent, potassium levels are invariably inadequate and inherent deficiencies of calcium and magnesium may be exacerbated by acidification processes which bind metal carbonates in neutralization reactions (Gemmell, 1977). Apart from situations where trace metals occur in phytotoxic concentrations, supplies of the essential micronutrients, boron, iron, manganese, copper, zinc and molybdenum are usually satisfactory (Murray, 1973). However, circumstances may occur in which specific trace elements are deficient. For example, at Bougainville Copper Mine, Papua New Guinea, boron deficiency has been identified as a limitation to plant growth (Hartley, 1976). Micronutrient availability is closely related to substrate pH (Fig. 6.3), and in particular to the nature of the gangue minerals and the problems of ion competition and antagonism.

TABLE 6.3

NON-FERROUS METALS IN SPOIL FROM ABANDONED MINES IN UK. (AFTER JOHNSON AND BRADSHAW, 1977)

Mining region	Counties	Number of sites surveyed	Concentration range[a]			
			Principal base metals			Associated metals[b]
			Cu	Pb	Zn	
Southwest England	Devon and Cornwall	16	65–6 140	48–2 070	26–1 090	Ag: <5–350, As: 68–7080, Cd: 5–145, Sn: 80–6 200
West and northwest England	Salop and Cheshire	12	15–7 260	840–26 000	980–21 000	As: 93–1 970, Ba: <100–14 000, Co: <5–95, Ni: 17–20
North Pennines	North Yorkshire and Durham	8	—	605–13 000	470–28 000	Ba: 400–62 000, Sr: 125–2 530, Cd: <2–325
South Pennines	Derbyshire	17	—	10 800–76 500	12 700–42 000	Ba: <100–104 000, Cd: <2–195, Cu: 29–240, Sr: <50–8 760

Lake District	Cumbria	7	77–3 800	2 070–6 370	4 690–7 370	Ba: <100–3 800
Central Wales	Powys and Dyfed	10	—	1 670–54 000	475–8 000	Ag: 8–100, Cd: 15–445, Cu: 77–560
North Wales	Clwyd and Gwynedd	19	30–5 750	6 400–76 000	11 300–127 000	Ag: 18–95, Ba: <100–7 500, Cd: 70–510, Ni: 30–695
South Scotland	Dumfries and Galloway	6	—	4 730–28 300	1 600–31 400	Cu: 10–680, Ni: <5–665
Range in normal, uncontaminated soils			2–100	2–200	10–300	Ag: <1, As: 1–50, Ba: 100–3 000, Cd: <1, Co: 1–40, Ni: 5–500, Sn: <10, Sr: 50–1 000

[a] Total values in μg/g of sieved (2 mm) air-dried spoil.
[b] Elements present in anomalous amounts at more than two sites. Elevated levels of Bi, Ge, Li, Mo and Sb were also found occasionally.

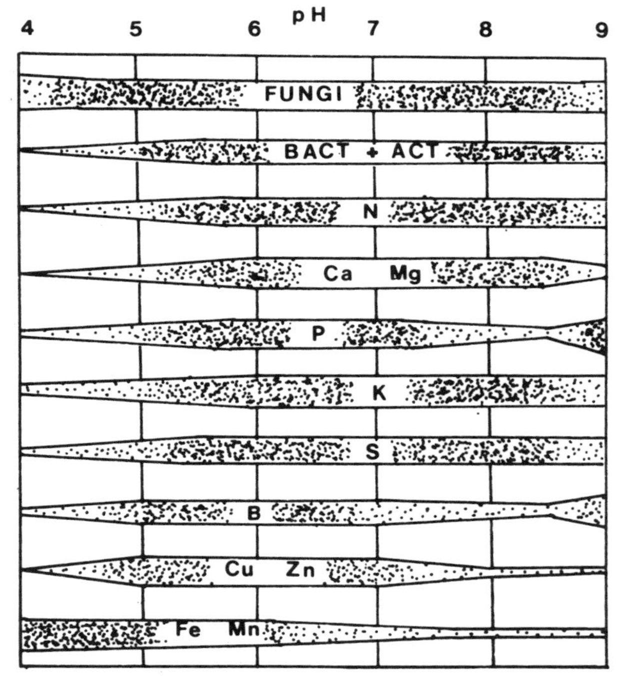

FIG. 6.3 The interactions between mineral soil pH, microorganism activity and availability of plant nutrients. (Murray, 1973, after Buckman and Brady, 1960.)

5.5. Climate

Climatic factors will greatly influence the system of reclamation that should be adopted. Metalliferous mines occur in all climatic regions of the world—including desert and permafrost zones where careful selection of well-adapted plant species is the basis of successful reclamation. Accurate information on the quantity and annual distribution of rainfall together with evaporation data is essential, for it enables broad-scale predictions of nutrient movement in spoil materials, leaching behaviour of heavy metal ions and the probability of salt crust formation.

6. REVEGETATION PROCEDURES

6.1. Amelioration Techniques

The successful revegetation of metalliferous mine wastes depends upon the amelioration of constraints to plant growth. Unfavourable textures can be improved by incorporation of organic matter such as straw, wood by-products (e.g. sawdust, chips), broiler litter, sewage sludge or farmyard manure. The latter organic residues have the added advantages that they provide essential nutrients, particularly nitrogen, in slow release form, and further benefit the reclamation scheme by complexing heavy metals and rendering them innocuous at least until such time as the organic matter degrades. Alternatively, on tailings dams, cereals or green crops may be grown for a short period and then incorporated into the surface to improve soil structure. This system has been adopted at the Cyprus Pima copper mine in Arizona (Ludeke, 1973).

Chemical amelioration of acidic wastes involves incorporation, or where this is impractical, surface treatment with agricultural lime or ground limestone to amend the pH conditions and reduce the availability of phytotoxic metals. In strongly acidic substrates as much as 40 t/ha of lime have been used, and where potential acidity must be countered more than double this figure may be required (Down and Stocks, 1977*a*; Murray, 1973). Under very low pH conditions liming treatments are not always effective. For example, applications of 12–13 t/ha of lime to certain gold mine tailings in South Africa have been found to give only a marginal and temporary improvement in pH conditions. Recommendations have been made that before sowing or planting can be considered, a minimum pH of 5·5 to 15 cm depth should be obtained (Chamber of Mines, 1979). Where this cannot be achieved by liming, spray irrigation has been advocated if the concentration of pyrite is below 0·2 % w/w. If pyrite levels exceed this threshold the natural processes of oxidation and weathering must be accelerated by surface ridging or ploughing before neutralization of acidity can be attempted.

Amelioration of the inherent low fertility involves application of inorganic fertilizers and/or organic matter, and in many situations the establishment of species capable of fixing atmospheric nitrogen. In Great Britain, 400–800 kg/ha of a high-phosphate, complete (NPK) fertilizer has been recommended as a primary treatment to rectify macronutrient deficiencies (Bradshaw and Chadwick, 1980). The deployment of largely insoluble fertilizers, for example basic slag and slow-release compounds such as magnesium ammonium phosphate, allows for the progressive release of plant-available nutrients over a longer period of time. This may be advantageous in temperate and wetter climates where leaching losses of nutrients can eventually lead to the onset of sward deterioration. If normal fertilizers are to be used, repeated treatments at a low rate are usually superior to a single high-rate application, particularly in terms of nitrogen recovery by the sward. The susceptibility of this nutrient to leaching losses indicates the importance of including nitrogen-fixing legumes in reclamation seed mixtures, even though they can prove difficult to establish in acidic conditions. Pre-treatment by scarification may be necessary for rapid germination, and inoculation with the appropriate symbiotic bacteria is necessary for effective nitrogen fixation.

Repeated maintenance applications of lime and fertilizer are nevertheless required in many mine waste reclamation schemes to avoid sward regression. Bradshaw and Chadwick (1980) recommend annual applications of 25 kg/ha of N, P and K for at least six years, whereas Down and Stocks (1977*a*) advocate a post-establishment treatment of 75 kg N/ha to counteract the considerable losses which occur in the early stages of sward development. Management programmes must, however, recognize the susceptibility of legumes to competition from co-established grasses under high-nitrogen regimes, and the adverse effect on symbiosis of regular maintenance treatments of nitrogen at high application rates. In South Africa the recommended fertilizer treatment for gold mine residues is 60 kg N/ha, 20 kg P/ha and 30 kg K/ha to be applied annually to sand dumps, biennially to the perimeter of slimes dams, and every third year to slimes dam surfaces (Chamber of Mines, 1979). In all cases, however, swards should be assessed individually on the colour of the vegetation, percentage ground cover, seeding potential, and the results of chemical analyses before decisions are taken on the type of fertilizer and rates of application for aftercare. Controlled grazing or, where this is undesirable, regular cutting of swards, may be necessary to encourage tillering of grasses, and in hot climates to minimize the fire hazard presented by prolific growth.

6.2. Overburden

Techniques for revegetation of overburden material, which may be used to cover waste rock dumps or even hostile tailings ponds, depend greatly on the way in which the material has been handled and stockpiled. The initial stripping programme should recognize horizon differentiation within the soil profile, and allow for separate removal and storage of topsoil, subsoil and deeper layers. This enables the stripping programme to be reversed during replacement operations. Compacted surfaces should be mechanically ripped before applying seed and fertilizer. A comprehensive aftercare policy will be necessary, particularly if topsoil replacement cannot be achieved.

6.3 Waste Rock

Recontouring of stockpiles and rock dumps to provide slopes of less than 15° is essential if vegetation establishment is to be achieved using traditional agricultural machinery (Kohnke, 1950). Hydroseeding techniques can be utilized to give access to steeper slopes and remote locations (Brooker, 1974). In either case the hostile physico-chemical characteristics of the material must be countered if a permanent and self-maintaining sward is to be established. This may involve applications of liming material and infilling of the larger voids with suitable amendments; or, in situations where this is impractical, surface treatment with 10–50 cm of overburden or other innocuous ameliorants. This approach also counters the extreme moisture stress that would otherwise occur. Incorporation of organic residues can help to rectify these problems, but regular maintenance applications of lime and compound inorganic fertilizer will probably be indispensable.

6.4. Tailings Dams

Within the last decade research into the problems of revegetating disused metalliferous tailings dams has expanded considerably. The increasing scale of modern dams emphasizes the importance of forward planning of reclamation schemes to suit the particular characteristics of the substrate, and the ultimate land-use objectives. The diversity of revegetation techniques is considerable and to some extent reflects the range of metals currently mined worldwide, as well as the age of different materials in relation to changes in processing technology. The very high metal content of older tailings, such as those discarded during the 'heyday' of metal mining in Britain (1750–1900), may demand the use of more specialized reclamation procedures than are necessary for recently abandoned dams from modern operations in the same orefields (Figs. 6.4 and 6.5). Equally, however, there are many examples where the constraints on revegetation are just as severe with modern tailings as with older wastes. This is particularly true with highly acidic tailings where the revegetation problems of both metal contamination and pyrite oxidation are accentuated by the obstacles of scale and inaccessibility to heavy machinery.

Several attempts have been made to revegetate metalliferous tailings for environmental control purposes and landscape rehabilitation. Many of the successful schemes have been preceded by detailed analytical studies, by small-scale bioassays under glasshouse conditions, and by extensive field trial programmes, with the intention of identifying the major limitations to

FIG. 6.4 Abandoned and unstable metalliferous mine waste, containing 0·8 % Zn and 0·3 % Pb (Parc Mine, near Llanwrst, North Wales). (Photo: A. J. Tollitt.)

FIG. 6.5 Parc Mine following vegetative stabilization by surface treatment with 30 cm of quarry overburden and seeding with a heavy metal (Pb/Zn) tolerant cultivar of *Festuca rubra*. (Photo: A. J. Tollitt.)

FIG. 6.6 Surface of a recently disused fluospar tailings dam containing 1 % Pb and 0·2 % Zn (La Porte Industries Ltd, Eyam, Derbyshire, England).

FIG. 6.7 The tailings dam 18 months after revegetation works were implemented (Johnson *et al.*, 1976).

TABLE 6.4

APPROACHES TO REVEGETATION

Waste characteristics	Reclamation technique	Problems encountered
Low toxicity Toxic metal content: <0·1%. No major acidity or alkalinity problems.	*Amelioration and direct seeding with agricultural or amenity grasses and legumes* Apply lime if pH < 6. Add organic matter if physical and chemical amelioration required. Otherwise apply nutrients as granular compound fertilizers. Seed using traditional agricultural or specialized techniques (e.g. Pb/Zn/CaF_2 tailings: Johnson *et al.*, 1976).	Probable commitment to a medium/long term maintenance programme. Grazing management must be strictly monitored and excluded in some situations.
Low toxicity and climatic limitations Toxic metal content: <0·1%. No major acidity or alkalinity problems. Extremes of temperature, rainfall, etc.	*Amelioration and direct seeding with native species* Seed or transplant ecologically adapted native species using amelioration treatments (e.g. lime, fertilizer) where appropriate (e.g. Cyprus–Pima Cu tailings: Ludeke, 1973).	Irrigation often necessary at establishment. Expertise required on the characteristics of native flora.

High toxicity Toxic metal content: >0·1%. High salinity.	*(1) Amelioration and direct seeding with tolerant ecotypes* Sow metal- and/or salt-tolerant ecotypes. Apply lime, fertilizer and organic matter, as necessary, before seeding (e.g. Pb/Zn waste: Smith and Bradshaw, 1972).	Possible commitment to regular fertilizer applications. Relatively few species have evolved tolerant populations. Grazing management not possible. Very few species are available commercially as tolerant varieties.
	(2) Surface treatment and seeding with agricultural or amenity grasses and legumes Amelioration with 10–50 cm of innocuous mineral waste (e.g. overburden) or organic material (e.g. sewage sludge). Apply lime and fertilizer as necessary (e.g. Pb/Zn waste: Johnson *et al.*, 1977).	Regression will occur if shallow depths of amendment are applied or if upward movement of metals occurs. Availability and transport costs may be limiting.
Extreme toxicity Very high toxic metal content. Intense salinity or acidity. Radioactivity hazards.	*Isolation* Surface treatment with 30–100 cm of innocuous barrier material (e.g. clay, overburden, unmineralized rock) and surface blinding with 10–30 cm of a suitable rooting medium (e.g. subsoil, sand). Apply lime and fertilizer as necessary (e.g. Captains Flat, Pb/Zn/Cu wastes (Australia): Craze, 1977).	Susceptibility to drought according to the nature and depth of amendments. High cost and potential limitations of material availability.

plant growth and appropriate remedial treatments. This approach has been adopted for lead–zinc contaminated fluorspar tailings in Derbyshire, England, where future mining proposals within the Peak District National Park are likely to proceed only if accompanied by comprehensive rehabilitation plans (Figs. 6.6 and 6.7). In this situation the principal restoration objectives are the stabilization of abandoned tailings dam surfaces, and landscape redevelopment which is consistent with the predominantly agricultural surrounding countryside. Elsewhere, very different approaches have been used, but towards similar objectives. The range of revegetation options, together with their limitations, is outlined in Table 6.4 from which the importance of species selection as well as physico-chemical amelioration is clearly apparent.

7. SELECTION OF PLANT MATERIAL

The success of metalliferous waste reclamation schemes depends on suitable amelioration of adverse growth conditions coupled with careful selection of plant material in line with the characteristics of the substrate, land use and management objectives. Particular attention must be given to overall climatic conditions, and to acidity, nutrient status and metal contamination problems, especially where significant amelioration is impractical. Subsidiary local factors must also be considered, such as pest and disease incidence and the recovery ability of different species when subjected to extreme conditions (e.g. fire, drought).

7.1. Ground Cover Vegetation

Revegetation using native species should be based on several criteria. The flora of inoperative sites should be examined, as the species composition of pioneer vegetation can indicate species adaptations and the evolution of tolerant ecotypes. If rapid stabilization is required, growth rates and habit must be considered as well as rooting characteristics which are important in binding the surface layers. For example, establishment of a ground cover of annual species may lead to seasonal erosion problems and subsequent re-establishment difficulties, so herbaceous and woody perennials are vital components of seeding and planting specifications. Species selection must consider the ease with which propagation can be carried out, the abundance and viability of seed and, where several species are involved, their competitiveness in a complex sward. The commercial availability of the chosen species should be established at an early stage so that collections of wild material may be made if necessary (Table 6.5).

TABLE 6.5
NATIVE SEED COLLECTION METHODS

Method	Considerations
Manual collection from vegetation	Often the only method available. Costly unless cheap labour is obtainable (Quilty, 1975; Farmer *et al.*, 1976)
Collection and spreading of seed-bearing shrubs, grasses, etc., on to land to be restored	Difficult to establish correct time for collection and may damage area of collection (Quilty, 1975)
Direct return of topsoil with its associated seed bank	Perhaps the best method if soil is available on site. High costs may otherwise limit this method (Tacey *et al.*, 1977)
Vacuuming of soil surfaces and plants at, or shortly after, seed dispersal	An effective method for which suitable machines are being developed (Sauer, 1975)

In extreme conditions, such as the short growing season and generally harsh climate of arctic and alpine regions, it is essential that the specialized adaptations of the native flora are recognized and utilized fully. Revegetation schemes in these areas therefore rely heavily on the 300 or more tundra species (Brown *et al.*, 1978). Introduction of species from similar climates has produced promising results in the semi-desert Pilbara region of north-west Australia, where the exotic grasses buffel (*Cenchrus ciliaris*), birdwood (*Cenchrus setigerus*) and kapok (*Aerua tomentosa*) have been successfully established (Martinick, 1977). Moreover, eucalypts (e.g. *Eucalyptus camaldulensis*), native to Australia, are now used in mine waste revegetation schemes throughout the world. Examples where intra-specific variation has been exploited include the Mt Isa mines in Australia where seed of native species has been collected from natural salt pans, mineralized outcrops, and old mining dumps for use in revegetating modern tailings dams and waste rock dumps (Ruschena *et al.*, 1974).

In Britain this principle has been developed further and three metal-tolerant grass cultivars are now undergoing commercial seed production for revegetation work in temperate climates (Table 6.6). However, there is a developing need for metal-tolerant material suited to toxic mine wastes in tropical, sub-tropical and arid regions where genera such as *Agrostis* and *Festuca* will be of limited practical value since they do not possess the necessary climatic adaptations.

Situations where mine waste revegetation has an agricultural objective are uncommon, but with continued improvements in processing technology further developments in this direction are very probable, particularly where edaphic conditions permit and where such land uses are consistent with those of the immediate locality. Moreover, agricultural grasses and legumes are often used in reclamation work even where grazing management does not constitute part of the redevelopment plan. In general

TABLE 6.6
METAL-TOLERANT CULTIVARS AVAILABLE ON A COMMERCIAL SCALE

Species	Cultivar	Application
Festuca rubra	Merlin	Neutral–calcareous Pb/Zn wastes
Agrostis tenuis	Goginan	Acidic Pb wastes
Agrostis tenuis	Parys	Acidic Cu wastes

terms, an agricultural endpoint will be very demanding of rehabilitation measures and possibly cost-prohibitive amelioration techniques will be required. Careful selection of seed mixtures, based on trial programmes, is again critical and has consistently proved more successful than seeding with a diverse mixture on the assumption that the best adapted will survive (Murray, 1973). Recommended seeding rates for revegetation work (45–135 kg/ha) greatly exceed standard agricultural rates (17–28 kg/ha) even when traditional cultivation techniques are employed (Down and Stocks, 1977*a*); but still higher values (80–250 kg/ha) have been advocated to compensate for desiccation and dispersion losses where specialized techniques are used (e.g. hydroseeding, aerial seeding).

7.2. Trees and Shrubs

The role of trees and shrubs in metalliferous waste reclamation relates closely to the characteristics of individual sites and the precise objectives of revegetation programmes (Table 6.7). Commercially orientated planting programmes based on coniferous and broadleaved species of economic value are rarely feasible, though a limited number of schemes have been implemented for production of timber, wood fire, pulpwood and associated materials. More commonly, woody species are established for landscaping purposes, protection against wind erosion, and to encourage the development of wildlife interest. Unfortunately, establishment failures and inferior performance in the long term are not uncommon on mine tailings and waste rock dumps. This often stems from a disregard of the need to

TABLE 6.7
THE VALUE OF TREES IN RECLAMATION

Objectives	Case history outlines
Supply local timber needs	In Nigeria, firewood and structural timber are needed by the local native community (Wimbush, 1963)
Provision of employment in rural areas	Employment provided by forestry helps to promote stable communities in areas where mining operations have ceased (NCAIOP, 1969)
Enhancement of scenic values	Strategic planting can help to break up geometric lines and artificial skylines created by mine wastes. Ornamental trees can be planted on the fringe of forestry to provide an attractive amenity feature (NCAIOP, 1969).
Establishment of perimeter wind breaks and shelter belts	At Hollinger gold mine in Canada, Carolina poplar—a fast-growing tree with a spreading root and branch system—has been used for perimeter wind breaks (Gordon, 1969)
Provision of food and shelter for wildlife	The International Nickel Company has redeveloped tailings and adjacent areas in Ontario, Canada, for wildlife management and is negotiating with the Ontario Ministry of Natural Resources to have the areas so designated (Peters, 1978)
Reduction of surface water and resulting erosion by increasing the evaporation potential	Trees increase the evaporation rate and with relatively dense canopies, large amounts of rain are prevented from reaching the ground, being evaporated directly. A single 20-year-old spruce, 6 m high, will transpire about 379 litres of water/day (Leroy and Keller, 1972)
Commercial forestry	To provide for future yields of high quality forest products, trees are often planted in areas where mature timber has been harvested (Dickinson and Youngman, 1971). In order to promote local industry, high value timbers (e.g. teak and mahogany) can be planted where conditions permit, such as at Groote Eylandt manganese mine, Australia (Langkamp and Dalling, 1977)

relate species selection and planting techniques to the growth limitations of the substrate.

On a world scale a wide range of species has been recommended for use in different situations. Particularly successful species are birch (*Betula pubescens*), austrian pine (*Pinus nigra*), virginia pine (*Pinus virginiana*) and sallow (*Salix cinerea*), all of which are tolerant of low fertility and sometimes occur as natural pioneers of abandoned mine workings. Nitrogen-fixing species such as common alder (*Alnus glutinosa*), black locust (*Robinia pseudoacacia*) and *Acacia saligna* (*A. cyanophylla*) have been equally successful in warm climates. However, truly heavy-metal-tolerant populations of these and other species are virtually unknown, so considerable importance is attached to planting methods and amelioration treatments.

Both waste rock dumps and tailings require treatment to ensure satisfactory establishment and growth in the long term. Planting into individual pockets or trenches backfilled with an organic amendment (topsoil or subsoil) is normally satisfactory, provided that post-establishment maintenance fertilizer treatments are applied. Except in the case of high nutrient status amendments (e.g. sewage sludge), initial fertilizer treatments range from 100–300 g of NPK fertilizer (17:17:17)/50 litres of amendment, with further annual additions of 50–200 kg/ha of N, P_2O_5 and K_2O for three to five years after establishment. In situations where physical reshaping and surface amelioration are impractical, unconventional approaches to tree and shrub establishment may overcome the adverse growth conditions. Injection of a peat–wood cellulose slurry mixed with fertilizer and limestone has successfully countered the drought problem resulting from voids in coarse non-metalliferous wastes and adaptations of this technique have considerable potential for tree establishment on waste rock stockpiles (Sheldon and Bradshaw, 1976).

The ideal transplant material will vary with the nature of the site and the landscaping objectives. Well-rooted nursery stock comprising young seedlings or transplants, whips or more advanced material is commonly used, but direct seeding with surface mulching is a viable alternative for inaccessible areas provided that appropriate seed pre-treatment measures have been carried out. For example, *Acacia* and *Robinia* seed should be scarified before use in large-scale reclamation schemes. The optimum approach to tree and shrub establishment will be related mainly to substrate and climatic conditions but also to other considerations; for example, the immediate visual benefits of transplanting relatively advanced stock must be equated with the increased risk of establishment failure and in this event the added costs of replacement planting.

8. CONCLUSIONS

Considerable attention has been focused worldwide on the environmental effects of modern non-ferrous metal mining operations during and subsequent to their active life because of their impact on terrestrial and aquatic ecosystems. Past experience suggests that one of the most significant consequences of metal mining in terms of environmental quality is the continued dispersal of heavy metals from abandoned workings and in particular from disused tailings dams, waste rock dumps and other mineralized residues. Effective means of preventing wind and water erosion of these materials are clearly necessary.

Stabilization by vegetation establishment is generally superior to alternative techniques and offers several advantages over physico-chemical methods. The potential exists for achieving stabilization which will be permanent and therefore remain effective beyond closure of the operation when site maintenance will be minimal or discontinued. Moreover, the considerable visual benefits to be gained from revegetation include amelioration of the intrusive character of disused operations which is so frequently emphasized when planning permissions are sought for new sites. Outline revegetation plans should be formulated at the planning stage of mining operations provided that the characteristics of the various wastes which will be produced are known. Detailed submissions can only be considered viable if they are based on experimental studies which confirm that the revegetation and final land use proposals are technically feasible. Chemical and physical analyses, glasshouse and field trial programmes can yield valuable data in this connection.

In theory, the range of revegetation options is considerable, but exclusion of certain possibilities may be necessary based on the climatic, substrate and various technical limitations pertaining to individual sites. The objective in selecting a particular approach to revegetation must be to meet existing environmental constraints most effectively and also most economically. Provided that the limitations to plant growth and development are thoroughly understood this should be possible within the financial constraints of individual mining operations.

REFERENCES

Andrews, R. D. (1975). Tailings: Environmental consequences and a review of control strategies. In: *Symposium Proceedings, Vol. III, International Conference on Heavy Metals in the Environment* (T. C. Hutchinson (ed)), Toronto, Canada, pp. 645–75.

Bell, A. V. (1974). The tailings pond as a waste treatment system. *C.I.M. Bulletin.* **67**: 73–8.

Bell, A. V. (1975). Base-metal mine waste management in Canada. In: *Minerals and the Environment* (M. J. Jones (ed)), Institute of Mining and Metallurgy, London, pp. 45–59.

Boorman, R. S. and D. M. Watson (1976). Chemical processes in abandoned sulphide tailings dumps and environmental implications for north-eastern New Brunswick. *C.I.M. Bulletin.* **69**(772): 86–96.

Bradshaw, A. D. and M. J. Chadwick (1980). *The Restoration of Land: The Ecology and Reclamation of Derelict and Degraded Land.* Blackwell Scientific Publications, Oxford, 309 pp.

Brooker, R. (1974). Hydraulic seeding techniques: An appraisal. *J. Inst. Landsc. Arch.* **108**: 30–2.

Brown, R. W., R. S. Johnston and K. van Cleve (1978). Rehabilitation problems in alpine and arctic regions. In: *Reclamation of Drastically Disturbed Lands* (F. W. Schaller and P. Sutton (eds)), Am. Soc. of Agronomy, Crop Sc. Soc. of Am., Soil Sc. Soc. of Am., Wisconsin, USA, pp. 23–44.

Buckman, H. O. and N. C. Brady (1960). *The Nature and Properties of Soils*, 6th Edition. Macmillan Co., New York, 567 pp.

Burton, T. A., G. F. Gifford and G. E. Hart (1978). An approach to the classification of Utah mine spoils and tailings based on surface hydrology and erosion. *Environ. Geol.* **2**: 269–79.

Chamber of Mines (1979). Handbook of guidelines for environmental protection. Vol. 2. *The Vegetation of Residue Deposits Against Water and Wind Erosion.* The Chamber of Mines of South Africa, Johannesburg.

Costigan, P. A. (1979). The reclamation of pyritic colliery waste. Ph.D. Thesis, University of Liverpool.

Craze, B. (1977). Restoration of Captains Flat Mining Area. *J. Soil Cons. Serv. N.S.W.* **33**(2): 98–105.

Dean, K. C. and R. Havens (1973). Comparative costs and methods for stabilization of tailings. In: *Tailings Disposal Today* (C. L. Aplin and G. O. Argall (eds)), Miller Freeman, San Francisco, pp. 450–74.

Dickinson, S. K. and D. G. Youngman (1971). Taconite tailing basin reclamation—A phase of multiple resource management. Proc. 32nd Annual Univ. of Minnesota Mining Symposium, pp. 137–42.

Down, C. G. (1974). Problems in vegetating metal-toxic mining wastes. In: *Minerals and the Environment* (M. J. Jones (ed)), Inst. of Mining and Metallurgy, London, pp. 395–408.

Down, C. G. and J. Stocks (1977*a*). *Environmental Impact of Mining*, Applied Science Publishers Ltd, London. 371 pp.

Down, C. G. and J. Stocks (1977*b*). Environmental problems of tailings disposal. *Mining Magazine.* **137**(1): 25–33.

Duncan, D. W. (1972). Environmental problems associated with the disposal of mining and milling waste. *Western Miner.* **45**: 14–24.

Farmer, E. E., B. Z. Richardson and R. W. Brown (1976). Revegetation of Acid Mining Wastes in Central Idaho, USDA, Forest and Range Experiment Station, INT-178, Utah.

Farrell, T. P. (1977). Rehabilitation of mine wastes in a monsoonal semi-arid environment. 1977 Environmental Workshop Rehabilitation Papers. Australian Mining Industry Council, Dickson, Australian Capital Territory.

Gemmell, R. P. (1977). *Colonization of Industrial Wasteland: Studies in Biology No. 80*. Edward Arnold, London, 75 pp.

Goodman, G. T., C. E. R. Pitcairn and R. P. Gemmell (1973). Ecological factors affecting growth on sites contaminated with heavy metals. In: *Ecology and Reclamation of Devastated Land*, Vol. 2 (R. J. Hutnik and G. Davis (eds)), Gordon and Breach, London, pp. 149–73.

Gordon, I. M. (1969). Erosion control at Hollinger mine tailings site. *Canadian Mining Journal*. **90**(6): 46–50.

Hartley, A. C. (1976). Tailings and waste, revegetation at Bougainville. Address to Papua New Guinea Scientific Society, Port Moresby.

James, A. L. (1966). Stabilizing mine dumps with vegetation. *Endeavour*. **25**: 154–7.

Johnson, M. S., A. D. Bradshaw and J. F. Handley (1976). Revegetation of metalliferous fluorspar mine tailings. *Trans. Inst. Mining Metall. (Sect. A: Mining Ind.)*. **85**: A32–7.

Johnson, M. S. and A. D. Bradshaw (1977). Prevention of heavy metal pollution from mine wastes by vegetative stabilization. *Trans. Inst. Mining Metall. (Sect. A: Mining Ind.)*. **86**: A47–55.

Johnson, M. S., T. McNeilly and P. D. Putwain (1977). Revegetation of metalliferous mine spoil contaminated by lead and zinc. *Environ. Pollut*. **12**: 261–77.

Kay, B. L. (1977). Hydroseeding and erosion control chemicals. In: *Reclamation and Use of Disturbed Land in the South-west* (J. L. Thames (ed)), University of Arizona Press, Tucson, Arizona, pp. 238–47.

Kohnke, H. (1950). The reclamation of coal mine spoils. *Adv. Agron*. **2**: 317–49.

Langkamp, P. and M. Dalling (1977). Mined land rehabilitation on Groote Eylandt—A study. *BHP Journal*. **1**: 42–9.

Leroy, J. C. and H. Keller (1972). How to reclaim mined areas, tailings ponds and dumps into valuable land. *World Mining*. **25**: 34–41.

Ludeke, K. L. (1973). Vegetative stabilization of tailings disposal berms of Pima Mining Company. In: *Tailings Disposal Today* (C. L. Aplin and G. O. Argall (eds)), Miller Freeman, San Francisco, pp. 606–14.

Martinick, W. (1977). Iron-ore mining in the Pilbara: Rehabilitation—A joint approach. 1977 Environmental Workshop Rehabilitation Papers. Australian Mining Industry Council, Dickson, Australian Capital Territory.

Middleton, B. A. (1979). Land use planning, bauxite mining operation, Weipa. In: *Management of Lands Affected by Mining*, Workshop (May/June 1978) (R. A. Rummery and K. M. W. Howes (eds)), Commonwealth Scientific and Industrial Research Organization, Division of Land Resources Management, Kalgoorlie, Western Australia, pp. 140–54.

Murray, D. R. (1973). Vegetation of mine waste embankments in Canada. Information Circular IC301. Dept. of Energy, Mines and Resources, Mines Branch, Ottawa, Canada.

Murray, D. R. (1977). Pit Slope Manual, Supplement 10-1, Reclamation by Vegetation: Vol. 1—Mine Waste Description and Case Histories. CANMET (Canada Centre for Mineral and Energy Technology), Report No. 77-31.

NCAIOP (1969). Ironstone and agriculture. National Council of Associated Iron Ore Producers, Kettering, Northamptonshire, UK.

Nielson, R. F. and H. B. Peterson (1972). Treatment of mine tailings to promote vegetative stabilization. Bull. Agric. Experiment Station, Vol. 485, Logan, Utah.

Peters, T. H. (1970). Using vegetation to stabilize mine tailings. *J. Soil Water Cons.* **25**: 65–6.

Peters, T. H. (1978). Inco Metals reclamation program. *CIM Bulletin.* **71**: 104–6.

Quilty, J. A. (1975). Guidelines for rehabilitation of tailings dumps and open cuts. *J. Soil Cons. Serv.* **31**: 95–107.

Ruschena, L. J., G. S. Stacey, G. D. Hunter and P. C. Whiteman (1974). Research into the revegetation of concentrator tailings dams at Mt Isa. Paper presented at regional meeting on Recent Technical and Social Advances in the North Australian Minerals Industry, Australian Institute of Mining and Metallurgy, Victoria, pp. 1–16.

Sauer, R. H. (1975). Vacuumed soil as a seed source for revegetating strip mine spoils. *Min. Congr. J.* **61**: 16–19.

Sheldon, J. C. and A. D. Bradshaw (1976). The reclamation of slate waste tips by tree planting. *J. Inst. Landsc. Arch.* **113**: 31–3.

Shirts, M. B. and J. H. Bilbrey (1976). Stabilizing methods for the reclamation of tailings ponds. Symposium of the Australian Institute of Mining and Metallurgy, Landscaping Seminar 13, pp. 9–18.

Smith, R. A. H. and A. D. Bradshaw (1972). Stabilization of toxic mine wastes by the use of tolerant plant populations. *Trans. Inst. Mining Metall.* (*Sect. A: Mining Ind.*). **81**: A230–7.

Swaine, D. J. (1955). The trace element content of soils. Tech. Commun. No. 48, Commonw. Bur. Soil Sci., London. 157 pp.

Tacey, W. H., D. P. Olsen and G. H. M. Watson (1977). Rehabilitation of mine wastes in a temperate environment. 1977 Environmental Workshop Rehabilitation Papers. Australian Mining Industry Council, Dickson, Australian Capital Territory.

Wimbush, S. H. (1963). Afforestation of restored tin-mining land in Nigeria. *Commonw. For. Rev.* **42**: 255–62.

CHAPTER 7

The Effect of Trace Elements on Lower Plants

K. J. Puckett and M. A. S. Burton
Atmospheric Chemistry Division, Atmospheric Environment Service, Environment Canada, Downsview, Ontario, Canada

1. INTRODUCTION

Recently, there has been an increasing amount of research into the accumulation of metals by cryptogams. High levels of metal accumulation by these plants have been described in a wide range of species from diverse habitats, but studies into the possible toxicity of these accumulations have been limited. Cryptogams may be subjected to high metal concentrations as the result of exposure to metal-rich man-effected or natural sources. Some examples of metal levels in these plants are given in Tables 7.1–7.4. The data listed are by no means exhaustive as extensive metal analyses have been completed as the result of biogeochemical (Lounamaa, 1956; Shacklette, 1967; Doyle *et al.*, 1973; Whitehead and Brooks, 1969) and air and aquatic monitoring surveys (Laaksovirta and Olkkonen, 1977; Pilegaard, 1978; Ruhling and Tyler, 1970; Ray and White, 1979).

There are a number of points which must be considered before discussing the effects of trace metals on cryptogams. These include the role of the metals of concern in normal growth and the question of metal deficiency. It has not been determined whether all the trace elements required for normal growth by higher plants are essential to cryptogams. There is no evidence, however, to suggest that the basic metabolic processes of cryptogams are fundamentally different from those of higher plants or non-lichenized fungi and algae. Additional information on trace metals essential to lichens can be extracted from data on the growth requirements of non-lichenized algae

TABLE 7.1

ELEVATED AND BACKGROUND CONCENTRATIONS OF VARIOUS TRACE METALS IN LICHENS

Element	Background value[a]	Reference	Elevated values[a]	Reference
Cadmium	0·05	Sergy, 1978	334	Nash, 1975
Chromium	0·5	Jaakkola *et al.*, 1966	70	Schutte, 1977
Copper	0·7	Scotter and Miltimore, 1973	5 000	Poelt and Huneck, 1968
Lead	0·5	Puckett, 1978	1 131	Lawrey and Hale, 1979
Mercury	1	Sherbin, 1979	2·5	Connor, 1979
Nickel	1	Tomassini *et al.*, 1976	310	Tomassini, 1976
Zinc	6	Puckett, 1978	3 500	Maquinay *et al.*, 1961

[a] All values in ppm (dry weight).

and fungi (Ross, 1975) or from data on the growth requirements of algae and fungi isolated from lichens.

Some information on mineral requirements can be inferred from the substrate specificity of different species. This topic of substrate specificity is well reviewed for lichens by Brodo (1973). Substrate characteristics other

TABLE 7.2

EXAMPLES OF ELEMENT CONTENT OF BRYOPHYTES

Element	Background values[a]	Reference	Elevated values[a]	Reference
Cadmium	0·4	Ward *et al.*, 1977	148	Burkitt *et al.*, 1972
Chromium	0·5	Pakarinen and Tolonen, 1976	937	Lee *et al.*, 1977
Copper	5	Rastorfer, 1974	211	Rejment-Grochowska, 1976
Lead	0·7	Pakarinen and Tolonen, 1976	16 670	Shimwell and Laurie, 1972
Mercury	0·08	Solberg and Selmer-Olsen, 1978	15	Wallin, 1976
Nickel	0·15	Shacklette, 1965	146	Roberts and Goodman, 1974
Zinc	21	Rastorfer, 1974	7 166	Goodman and Roberts, 1971

[a] Values in ppm (dry weight).

TABLE 7.3
EXAMPLES OF ELEMENT CONTENT OF FERNS

Element	Background values[a]	Reference	Elevated values[a]	Reference
Copper	7	Hunter, 1953	455	Malaisse *et al.*, 1979
Mercury	0·2[b]	Sherbin, 1979	0·5[b]	Sherbin, 1979
Tin	0·5	Hunter, 1953	32	Peterson *et al.*, 1976
Zinc	21	Hunter, 1953	61	Peterson *et al.*, 1976
Cobalt	0·17	Hunter, 1953	196	Malaisse *et al.*, 1979

[a] Values in ppm (dry weight).
[b] Values in ppm (wet weight).

than chemical composition may dictate specificity, but there are a number of examples of lichens and mosses being restricted to one rock type. For example, *Rhizocarpon sphaericum* is restricted to serpentine rock (Ritter-Studnicka and Klement, 1968) and *Caloplaca heppiana* has a preference for calcareous substrates (Syers *et al.*, 1967). Wirth (1972) however, in discussing lichens characteristic of serpentine rock, indicates that the lichens are present not as a result of the high metal requirements but as a consequence of more successful competition in an unfavourable habitat, and Richards (in Shacklette, 1961) stated that '... any factor checking growth of vascular plants favours moss vegetation'. An example of this is suggested by the rapid and successful colonization of copper-rich substrates by *Pohlia nutans* (Beschel, 1959). The apparently restricted distribution of the 'copper mosses', particularly the genera *Merceya* and *Mielichhoferia* to mineralized substrates has been reviewed by Persson

TABLE 7.4
EXAMPLES OF ELEMENT CONTENT OF *Equisetum*

Element	Background values[a]	Reference	Elevated values[a]	Reference
Chromium	0·2	Cannon *et al.*, 1968	0·7	Cannon *et al.*, 1968
Copper	1·5	Cannon *et al.*, 1968	184	Ray and White, 1976
Lead	0·5	Cannon *et al.*, 1968	47	Ray and White, 1976
Mercury	0·01[b]	Sherbin, 1979	0·4	Sherbin, 1979
Nickel	0·3	Cannon *et al.*, 1968	11	Cannon *et al.*, 1968
Zinc	4	Cannon *et al.*, 1968	5400	Robinson *et al.*, 1947

[a] Values as ppm (dry weight).
[b] Value as ppm (wet weight).

(1956) and Shacklette (1967). A requirement for copper by these bryophytes has been the subject of some speculation; other factors which have been considered include the presence of sulphur and the low pH associated with the habitats of these genera (Schatz, 1955).

2. ECOLOGICAL ASPECTS OF TRACE METAL INTERACTIONS WITH LICHENS AND BRYOPHYTES

The limited information on the effects of trace metals on bryophytes and lichens, both in the field and laboratory situation, is in part due to difficulties in discerning effects. A distinction must be made between acute and chronic effects; acute effects being the result of short-term, high-concentration exposure to metals which results in the death of tissues or of the whole plant, whereas chronic effects are usually a consequence of long-term, low-concentration exposure and affect growth or yield. Detection of the latter category of effects is especially difficult because of the characteristically slow growth rate of bryophytes and lichens.

The chemistry of the substrate can affect lichen morphology. Both Massé (1964) and Wetmore (1970) demonstrated arrested development of lichen thalli which is probably due to certain precursors, possibly trace metals, which are missing from or limiting in the substrates in question. The only report of trace metals definitely affecting lichen morphology is that of Lambinon *et al.* (1964) who described morphological changes in lichens found growing on substrates high in zinc. Some lichens, e.g. *Diploschistes scruposus* accumulated very high concentrations of zinc without undergoing appreciable morphological change. In contrast, other lichens such as *Cladonia cornutoradiata* and *C. pyxidata* had very high zinc contents but also showed appreciable morphological deformation. For these lichens, the podetia remained small and were also swollen, twisted, darkened and abnormally granular-squamous.

Changes in community structure both in terms of the number and abundance of lichen species are amongst the effects induced by metals. For example, the spatial distribution of lichens relative to attached and detached lengths of wire fencing wound around trees has been examined by Seaward (1974). Large areas of trunk surface were devoid of epiphytic cover and these areas coincided with wire contact, forming a zone slightly above and extensively below the points of trunk–metal contact. Analyses of moribund samples from the edge of the lichen 'desert' showed three to five-fold increases in the iron, lead and zinc content over that of suitable control samples; the metals being made available to the lichens as the result of leaching from the wire. Seaward suggests that all three metals acting in conjunction are responsible for the observed community changes.

A similar situation has been described by Beckett and Wainwright (1980), where thalli of *Parmelia saxatilis*, located on rocks directly below natural drip points from a galvanized fence, were moribund, and these changes had been induced over a relatively short period (5 months). Only zinc and potassium levels showed significant differences between normal and moribund tissue with moribund tissue showing a five-fold increase in zinc content and a 50 % reduction in potassium concentration. The authors concluded that the moribund states of the thalli could be attributed to the elevated zinc levels and that the potassium decrease was evidence of zinc-induced damage to membranes similar to that described elsewhere for other metals (Puckett, 1976). Other lichen communities have been shown to be affected by leachate from a zinc fixture but not changed to the same extent by run-off from an iron fixture, illustrating the relative toxicity of the two metals (Garty *et al.*, 1979). Moss and lichen species growing inside galvanized zinc cages became discoloured and subsequently died, whereas outside the area of the zinc netting the same species remained healthy (Harris, 1946). Leachates from lead inserts on tombstones can also induce lichen mortality (Fig. 7.1). These effects described above are localized but they do serve to demonstrate that lichen and moss communities can be altered as the result of metal toxicity. More extensive examination of such well-defined situations could yield useful information on dose–effect relationships and temporal aspects of metal toxicity.

Although some reports have mentioned 'moss deserts' or greatly reduced species composition in the vicinity of metal emission sources, few reports have tried to quantify this observation. Folkeson (1979*a*) has reported distribution of mosses in relation to distance from a brass foundry emitting copper and zinc, but little sulphur dioxide. Frequency and cover of the species at sites in coniferous woodland in the surrounding area were assessed. *Pohlia nutans* had the highest cover within 1 km of the foundry. Frequency and cover of *Pohlia* decreased with increasing distance from the foundry, in contrast to other species. Competition was considered to be a major factor; *Pohlia* clearly could tolerate the high metal concentrations but may have been unable to compete well in habitats where other species could grow successfully. In addition, Folkeson (1979*a*) noted lichens were affected; the frequency of *Cladina rangiferina* was reduced close to the foundry while other species of *Cladonia* showed higher frequencies in the vicinity of the foundry than further away.

FIG. 7.1 Changes in lichen cover on tombstones as the result of leachates from lead inserts. (Photograph—J. Laundon.)

Community structure may also be altered by changes in bryophyte and lichen fertility and growth. LeBlanc and De Sloover (1970) observed reductions in fertility in mosses and lichens in their study of the distribution of these plants in Montreal and linked the reduced fertility to poor air quality. The 'copper mosses' have never been recorded with sporophytes on copper-bearing substrates though fertile material has been collected from hot springs and from limonite (Noguchi and Furuta, 1956). *Pohlia nutans* growing in a copper swamp was sterile in the areas with high copper concentrations, and sporophytes were only observed at peripheral locations (Beschel, 1959). Also, interferences with germination and subsequent growth of vegetative propagules or spores can influence community structure. For example, spore germination of *Graphis* and *Lecidea* species was considerably reduced by treatment with copper or mercury solutions (Pyatt, 1976). Spore germination and protonemal development of *Funaria hygrometrica* were severely reduced by copper and the atypical 'capsule' cells formed at the higher copper concentrations were suggested as sites of copper immobilization by Coombes and Lepp (1974). Zinc was considerably less toxic than copper to both spore germination of *Funaria* and to the development of *Marchantia polymorpha* gemmae. At low concentrations of cadmium, some enhancement of development was noted for *Funaria* and *Marchantia* but at higher concentrations, *Funaria* spore germination was inhibited at concentrations which had no effect on *Marchantia* gemma development (Lepp and Roberts, 1977). Spore germination of two *Merceya* species was compared by Noguchi and Furuta (1956) who showed that *M. gedeana* from a habitat with elevated copper concentrations was more successful in both spore germination and regeneration from single leaves than *M. ligulata* from a non-mineralized soil. Growth rates were significantly suppressed in juvenile thalli of the lichen *Pseudoparmelia baltimorensis* subjected to a high atmospheric lead burden when compared to thalli exposed to lower lead conditions (Lawrey and Hale, 1979). However, larger thalli did not show this response in that there were no significant changes in growth response as the result of stress induced by the automobile exhaust emissions. Lead values in thalli close to the highway were much higher than in thalli further away, but it should be noted that lead is only one component of automobile exhaust and the observed growth reductions may well be a reflection of combined toxicities. Similarly, Rejment-Grochowska (1976) reported inhibition of growth of mosses transplanted for one year to urban sites at busy highway intersections. Growth resumed when the mosses were transplanted back to a relatively unpolluted site.

In areas subject to more widespread emissions from industrial and domestic sources, more obvious effects have been observed in terms of changes in cryptogam community structure. Coppins (1978), Crespo *et al.* (1977) and Lerond (1978) have all described major changes in lichen community structure in Edinburgh, Madrid and Rouen, respectively; all the changes have been linked with varying degrees of air quality. In this context, LeBlanc *et al.* (1974) considered frequency coverage of bryophytes on one habitat type around a smelter in Murdochville, Quebec. The moss species considered occurred at a distance of 6–7 km from the source, but a few did not occur within 20 km from the source. The survey indicated a better correlation of the index of atmospheric purity with the mineral contents of mosses—which intercept airborne pollutants over their entire surface—than of higher plants. Measurements of seven trace elements were made, but direct effects of these on species distribution could not be evaluated due to the presence of other pollutants in the area. In many instances, causal relationships linking such community changes to metal toxicity are hampered by the concurrent presence of other potentially toxic agents such as sulphur dioxide. Other examples of this situation include the simultaneous exposure of lichens to arsenic and sulphur dioxide in Yellowknife (Hocking *et al.*, 1978) and lichen and moss paucity in areas of high copper, nickel and sulphur dioxide emissions (Hocking and Blauel, 1977; Goodman and Roberts, 1971). In such cases, determining the contribution of the various metals to the observed effects is difficult. Also, these situations are complicated in that the existence of various pollutants may result in synergistic, additional or antagonistic effects.

The consequences of metal effects on lichens may not be restricted to the lichen thallus but may extend to the whole plant community. Lichens can be an important source of available nitrogen as the result of nitrogen-fixation by the blue–green algal component of certain lichens. Lichenized blue–green algae are among the principal agents of nitrogen fixation in grasslands, tundra and sub-arctic forests (Alexander, 1974; Huss-Danell, 1977). Reduction in nitrogen fixation as the result of man-induced metal emissions would be critical in situations where soil nitrogen deficiency is frequently the factor that ultimately limits primary productivity (Bliss, 1962; Haag, 1974). The nitrogen-fixing terricolous lichens *Collema tenax* and *Lecidea* sp. are abundant in the grasslands area of Montana. Nitrogen fixation by these two species was reduced by a mixture of sodium fluoride and lead chloride simulating the ratio for the predicted emissions for these elements in this area (Sheridan, 1979). In a separate study, nitrogen fixation by the phycobiont of *C. tenax* was generally stimulated by low concentrations (0·005–0·025 ppm) of arsenic, cadmium, lead, nickel, palladium and zinc but with the exception of arsenic and palladium, higher concentrations of these metals (0·025–0·125 ppm) inhibited fixation (Henriksson and DaSilva, 1978).

3. PHYSIOLOGICAL ASPECTS OF TRACE-METAL INTERACTIONS WITH LICHENS, BRYOPHYTES AND FERNS

The physiology of toxicity begins with an increased metal supply to the plant and proceeds to the failure of an 'essential to life' plant process (Foy *et al.*, 1978). Metal accumulation may be the first stage in the physiology of toxicity. The metals are accumulated by several mechanisms: particulate entrapment, ion exchange, electrolyte sorption and processes mediated by metabolic energy (Nieboer *et al.*, 1978). The accumulation of particulates, if the particles are not readily solubilized, may be of little metabolic significance and the particulates would be much less toxic when compared to the effects induced by an equivalent amount of metal in solution. Thus, for a full appreciation of metal toxicity, some understanding of the uptake mechanisms of metals from dilute solution is required.

Metal ion uptake from aqueous solutions by bryophytes and lichens has received considerable attention. Laboratory studies on lichens by Tuominen (1967), Handley and Overstreet (1968), Puckett *et al.* (1973), Wainwright and Beckett (1975), Brown (1976), Nieboer *et al.* (1976*a*,*b*) and Burton *et al.* (1981) allow for a detailed description of short-term uptake mechanisms. The process is dependent on pH, concentration and time but independent of light, oxygen and temperature for several metals. Uptake is stoichiometric and is not reduced by metabolic inhibitors. Essentially, initial metal ion uptake takes place by an ion-exchange mechanism and the metal is accumulated on exchange sites on the cell wall. There is little evidence for intracellular uptake of trace metals but this may be as a consequence of the short-term nature of the experiments thus far. The available evidence on metal uptake by bryophytes also suggests that the major part of uptake is by ion exchange onto the cell wall sites (Ruhling and Tyler, 1970; Brown and Bates, 1972). Pickering and Puia (1969), however, found evidence for a metabolic component in addition to ion exchange for zinc uptake by *Fontinalis antipyretica*.

Studies of metal toxicity on bryophytes have mainly used the induction of plasmolysis as a criterion for injury. Interest in the effects of copper, stimulated by the descriptions of the 'copper mosses' (see Section 2), led to experimental work on these species. Url (1956) collected mosses from an old copper mining area and investigated the effects of metals on plasmolysis of the leaf cells. Cells were described as having resistance to the metal if they were not plasmolyzed during exposure and remained alive (as tested by subsequent plasmolysis in potassium nitrate or glucose). *Mielichhoferia elongata* and *M. nitida* both exhibited a high resistance to copper, all cells remaining alive in concentrations up to 5×10^{-2} M, and nearly all cells of *M. elongata* were alive in 5×10^{-1} M solutions (48-h exposures). After culture on nutrient agar for two months, there was some reduction in the resistance to high concentrations (1×10^{-1} M), but the resistance of some other species increased after culture. A surprising response by two liverworts was the observation of 'death zones', where, with a series of copper solutions of increasing concentration, cells were initially all living, then at higher concentrations some or all were dead, but at still higher concentrations, the cells were all or mostly alive. Two of the species from the copper mine showed this response, as did two others from a different locality. Beschel (1959) also reported such 'death zones' for *Pohlia nutans* from a copper swamp. Url (1956) suggested that copper resistance was due to the formation of a protective 'layer' at the plasmalemma surface at the high concentrations which would prevent further entry of the metal.

Intermediate concentrations would be too high for the plasma to tolerate, but too low for the 'layer' formation and would cause death of the cells.

The response of several bryophyte species to various metals had previously been reported in terms of 'plasmatic resistance' by Biebl (1947, 1950*a*,*b*). This method had been used in 1924 by Pringsheim (in Biebl and Rossi-Pillhofer, 1955) to demonstrate the survival of *Mnium rostratum* cells in extremely high (18–20 %) zinc sulphate solutions. Considerable variation between species was found, and Biebl (1947, 1950*a*) indicated that in some species, resistance to zinc and boron also showed seasonal variations. Leaf border and mid-rib cells of *Mnium rostratum* were more sensitive to zinc than the lamina cells, but less sensitive to boron. Also, younger leaves were more sensitive than older leaves to zinc, though not to boron.

Plasmatic resistance to vanadium by mosses and liverworts was found to be much higher than that of higher plants (Biebl, 1950*b*). Some species were resistant to up to 20 % vanadium sulphate, whereas Url (1956), while finding resistance to vanadium was greater than to other metals, recorded 'death zones' as also described for copper. *Alicularia scalaris* cells for example were dead in 5×10^{-7} M vanadium sulphate, most were alive in 5×10^{-3} M and 1×10^{-1} M, and above this, again most cells were dead. Resistance to chromium was reported by Url (1956) to be highest in one of the 'copper mosses', *Mielichhoferia elongata*. These experiments on plasmatic resistance are based on the concept that plasmolysis results from penetration through the membrane, thus species having high resistance would be more likely to tolerate high external concentrations in their habitat, apparently by exclusion of the metal from the cells.

Photosynthesis and respiration in lichens have both been shown to be sensitive to excess metals. Nash (1975) watered *Cladonia uncialis* and *Lasallia papulosa* with solutions of cadmium (20 μM–20 mM) and zinc (200 μM–20 mM) and after nine days the lichen photosynthetic and respiratory rates were monitored. Both photosynthesis and respiration were depressed after treatment with cadmium or zinc, with photosynthesis being reduced at much lower concentrations of these metals than respiration. The response of lichen photosynthesis to various metals has also been studied by Puckett (1976) and Richardson *et al.* (1979). Puckett (1976) examined the effect of various metals on photosynthetic ^{14}C fixation by *Umbilicaria muhlenbergii* and *Stereocaulon paschale*. The relative toxicities of these metals on photosynthesis is reflected in the series Ag, Hg > Co > Cu, Cd > Pb, Ni for short-term exposures and Ag, Hg > Cu ≥ Pb, Co > Ni for extended exposures. Richardson *et al.* (1979) also studied the photosynthetic response of *U. muhlenbergii* to several metals. Of the metals investigated only copper greatly reduced fixation, while lead and nickel had no effect under the experimental conditions used. Schutte (1976) showed that both photosynthesis and respiration by *Parmelia caperata* were affected after spraying the thalli with chromium. Initially, photosynthesis was not affected, but increasing the application rate of chromium, from four to six applications/4-week period, resulted in a 59% decrease in photosynthesis. Respiration rates were increased by 70% for the initial treatment but were reduced to 60% of the control when the chromium applications were increased. Johnsen (1975) showed that lichen respiration was not affected by low concentrations of cadmium, lead or vanadium (0·1–10 ppm). Exposure to higher lead concentrations than those used by Johnsen resulted in marked reductions in respiration by *Usnea fragilescens* and *Ramalina fastigiata* (Baddeley, Ferry and Finegan in James, 1973). Copper, mercury and nickel also reduced respiration by these lichens to a large extent.

Membrane functioning in plants is often impaired in the presence of pollutants (Chimiklis and Heath, 1975; Puckett *et al.*, 1977). This response is not unexpected in that plant membranes are the first metabolic entities that would come into contact with the pollutant. Such interactions may be significant in the event that induced alterations in membrane function could result in modifications in intracellular processes. Membrane functioning can be assessed by monitoring potassium efflux and studies on lichens and bryophytes have shown that membrane permeability can be altered by metals. Brown and Slingsby (1972) showed that low concentrations of nickel induced only slight losses of potassium from *Cladonia rangiformis* but that there was an abrupt increase in potassium loss at higher nickel concentrations. Nickel and cobalt produced a similar response by *Umbilicaria muhlenbergii* (Puckett, 1976). The high nickel and cobalt concentrations required to produce the large potassium efflux have little ecological relevance but nevertheless serve to demonstrate the lack of toxicity with respect to membrane integrity. In contrast, large potassium loss occurred in low concentrations of copper, mercury and silver, and increased with increasing concentration of metal in solution (Puckett, 1976). The response to copper, nickel and silver described above has also been shown for *Cladina rangiferina*. In addition, the effect of thallium was similar to that of copper while that of manganese resembled the response to nickel (Burton *et al.*, 1981). Incubation of *Solenostoma crenulatum* in copper solutions resulted in leakage of potassium ions, but samples from a copper mine were unaffected by concentrations which caused potassium loss from a lead mine population (Brown and House, 1978). At high concentrations, the copper mine samples also suffered membrane damage as shown by a sudden increase in potassium loss.

Laboratory studies have demonstrated some synergistic, additional or antagonistic reactions mentioned above. Sulphur dioxide in the presence of copper or nickel had a much greater effect in inducing a potassium efflux from *Cladina rangiferina* than did sulphur dioxide alone (Puckett, unpublished data). In contrast, iron and manganese are antagonistic to sulphur dioxide, in that the effect of this gas is prevented in the presence of the metals (Puckett, unpublished data). In experiments with *Umbilicaria muhlenbergii*, Nieboer *et al.* (1979) also observed that copper and lead increased sulphur-dioxide-induced potassium loss, while nickel and zinc conferred slight protection from the effects of sulphur dioxide. Examining the response of photosynthetic ^{14}C fixation to sulphur dioxide showed that lead or nickel had no influence on sulphur dioxide phytotoxicity (Richardson *et al.*, 1979). Lead did not modify the effects of sulphur dioxide on *Hypogymnia physodes* or *Lecanora conizaeoides* (Johnsen, 1975). Nash (1975) observed that when *Lasallia papulosa* was exposed to zinc and cadmium concurrently, the effect on photosynthesis was less severe than from each metal alone.

In the field, situations arise where such interactions could take place. After rain or dew formation, the lichen surface will be covered with a water film which could act as an effective sink for sulphur dioxide, since this gas is very soluble in water (Hocking and Hocking, 1977). This thalline water will contain dissolved metals originating from dry and wet deposition and will be in contact with particulates and with metals bound on ion-exchange sites. Metal concentrations in lichen thalli are in sufficient quantities to result in the modification of the sulphur dioxide effects (Puckett, 1978) but the extent of participation of these metals will depend upon their chemical form. Hence, under these conditions the metal content of lichens may play a role in determining the extent of injury due to sulphur dioxide.

Chlorophyll spectra can also be altered by various metals. No changes were noted in the chlorophyll absorption spectrum of *Umbilicaria muhlenbergii* exposed to cupric ions for 1–6 h, but incubation for 15 h produced marked deviations from the control spectra (Puckett, 1976). Two changes were noted, of which the most noticeable was a shift (10 nm) of both the red and blue peaks to shorter wavelengths. The other change was a loss in absorbance (bleaching) of the same chlorophyll peaks. This response to copper was also noted in species of *Peltigera*, *Dermatocarpon*, *Hypogymnia* and *Lobaria* (Puckett, 1976). Exposure of *U. muhlenbergii* to mercury and silver also produced spectral shifts and loss of absorption, but cadmium, cobalt, lead and nickel had no effect. Such changes in absorption spectra could involve processes associated with the allomerization of chlorophyll (Johnston and Watson, 1956; Holt, 1958, Seeley, 1966). Thalli of *Parmelia caperata* became chlorotic after spraying with beryllium (4·5 $\mu g\,ml^{-1}$), cerium (70 $\mu g\,ml^{-1}$) and chromium (26 $\mu g\,ml^{-1}$), and the chlorophyll content was reduced by 57%, 27% and 46%, respectively (Schutte, 1976).

Lichen thalli have been shown to exhibit extracellular enzyme activity. Boissière (1973) has demonstrated that some phosphatase activity is localized in the plasmalemma of the phycobiont (*Nostoc*) of *Peltigera canina*. Their extracellular location would make these enzymes particularly sensitive to changes in the immediate environment. Indeed, phosphatase activity has been used previously to measure effects on lichens both in the field (Bauer and Kreeb, 1974) and laboratory (Schmid and Kreeb, 1975). In the field surveys, the multiplicity of sources precluded the identification of any trace elements as agents responsible for the observed reductions in activity. Laboratory studies have shown, however, that several metals reduce phosphatase activity in whole thalli of *Cladina rangiferina*. Of the cations studied, the uranyl and vanadyl ions reduced activity by 60% and 86%, respectively. The anions biselenite, molybdate and vanadate also resulted in marked reductions (51%, 94% and 88%) in the phosphatase response (Lane and Puckett, 1979).

In the studies described above, the response to vanadium has special significance. Vanadium levels in atmospheric aerosols have been increasing steadily as the result of increasing fossil fuel combustion (Salmon *et al.*, 1978). High vanadium values have been found in atmospheric particulates collected in the north-eastern United States (Gordon *et al.*, 1974; Barry *et al.*, 1975) and in certain areas within Canada (Barrie, 1978). Soluble vanadium concentrations of up to 0·12 ppm have been measured in the snow pack around a point-source in northern Alberta (Barrie & Kovalick, 1978). Laboratory studies have indicated that such levels can have a significant detrimental effect on the surface phosphatase activity of some lichens. Exposure to 0·2 ppm vanadium (either as vanadyl or vanadate) under conditions approximating the field situation, resulted in a reduction of the phosphatase activity shown by *Cladina rangiferina* (LeSueur and Puckett, 1980). The actual field conditions may enhance or ameliorate this effect. In spring, 50–80% of the metals within the snow pack may be lost within the first 30% of the meltwater, resulting in metal concentrations in this initial meltwater being 2·5 times greater than that in the snow pack itself (Johannessen and Henriksen, 1978). Other elements present in the meltwater might influence the enzymatic response. High concentrations of aluminium, nickel or sodium and other metals have little effect on the action of the lichen phosphatase system (Lane and Puckett, 1979) when tested alone, but no information is available on the effects of mixtures of these metals on the enzyme response. There is also no information as to whether phosphatase activity can return after a suitable recovery period. The relevance of these laboratory studies to the field situation requires careful consideration. For example, what is the significance of the observed reductions in photosynthesis, enzyme activity and respiration with respect to growth? In addition, there is no information on the recovery of lichens or bryophytes with respect to any aspect of metal toxicity. Apparently normal growth may continue despite intermittent reductions in critical physiological processes.

Threshold values for metal-induced damage to lichens have received some attention (Nash, 1975). The demonstration of high metal accumulation by lichens under field conditions may have little relevance to an assessment of the phytotoxicity of these metals. A considerable proportion of metal accumulated by lichens may be as the result of particulate entrapment. Such particulates, if not readily solubilized, may be of little metabolic significance, yet they contribute to the total metal content and 'toxic' threshold levels are sometimes based on this total value. Also, the importance of determining the metal dose to the lichen in terms of concentrations supplied per unit time is illustrated by Nash (1975). Nash comments that the experimental data involving *Lasallia papulosa* at first appears to be inconsistent with field data. In the field situation, this lichen had zinc concentrations of up to 2560 ppm, yet net photosynthesis in laboratory experiments was depressed below zero by zinc concentrations of 482 ppm. The former measurement, however, represented total accumulation over 70 years but the latter over only nine days. Robitaille (in Rao *et al.*. 1977) suggests threshold limits for cadmium, copper, lead and zinc on the basis of lichen chlorophyll levels in the field. Near to the smelter studied, there were higher metal levels and correspondingly lower chlorophyll contents. However, the threshold values quoted again represent total accumulation which is integrated over an extended time period. In this instance, conclusions drawn about the phytotoxicity of the metals concerned should be considered in the light of the accompanying sulphur dioxide concentrations.

One effect of increased atmospheric levels of metals has been to elevate metal levels in lichens and mosses over large geographical areas (Steinnes, 1977; Groet, 1976; Pilegaard, 1978). Such surveys have been extended to include temporal as well as spatial aspects of metal accumulation. Rasmussen (1977) collected epiphytic bryophytes from two tree species in rural Denmark and demonstrated an increase in metal content over the past 25 years for all metals investigated, except for cadmium which remained almost constant. The highest increase noted was for vanadium (108 %) and the lowest increase was for nickel (42 %). In the period 1973–5, copper, iron and vanadium were increased by 24 %, 28 % and 32 %, respectively. Also, examination of herbarium material collected from the same locality in Montreal at various time intervals showed a general increase in metal levels from 1905/6 to 1977 (LeBlanc *et al.*, 1974). In these retrospective studies, data on the increasing metal levels were not accompanied by observations of any changes in appearance or cover of the moss concerned or any changes in the species composition of the community being studied.

Another approach to the estimation of metal loading is that described by Ruhling and Tyler (1970). They demonstrated that copper, iron and nickel were retained in almost unexchangeable form even when high concentrations of other ions were supplied, and showed a series for sorption and retention independent of whether single ions or a mixture of ions were in solution: Cu, Pb > Ni > Co > Zn, Mn. Such retention of metals which would be supplied in precipitation (and from dry deposition) allowed calculation of deposition rates of metals as measured in the three youngest segments of *Sphagnum* shoots (Ruhling and Tyler, 1971). Similarly, Groet (1976) has calculated deposition rates to bryophyte carpets but no information exists as to any effects resulting from these input rates. For example, at these measured input rates, will a decrease in growth result from critical values being exceeded? A temporal trend towards increasing lead values in lichens has been noted, together with reduced growth (Lawrey and Hale, 1979) but a threshold value for lead loading has not been determined. Some estimate of a threshold value for the effect of lead on the growth rate of *Rhytidiadelphus squarrosus* can be inferred from the data of Skaar *et al.* (1973). Cultivation of this moss for two months, with immersion in a series of lead solutions once in three days, demonstrated that applying concentrations of up to 50 ppm lead caused no significant effects on shoot elongation. When 100 ppm lead was applied, 50 % of shoots showed no growth, and at higher concentrations very little or no growth occurred in any shoots. The aquatic fern *Salvinia natans*, when cultured for four weeks in solutions containing up to 50 ppm lead, showed reduction in the number of fronds produced at concentrations above 10 ppm (Burton, unpublished data). Simola (1976) reported no effects of 1×10^{-4} M and lower concentrations of lead nitrate on the growth and cell structure of *Sphagnum fimbriatum*, although some degeneration of cell structure occurred at 1×10^{-4} M in *S. nemoreum*. Growth of both species was reduced by 1×10^{-3} M lead nitrate solutions. Similar experiments were reported for effects of cadmium, copper and mercury on *Sphagnum* growth (Simola, 1977). Investigations concerned with effects of metals on growth of *Salvinia natans*, showed that cadmium in concentrations greater than 0·1 ppm caused severe repression of growth, and death at 0·1 ppm (Hutchinson and Czyrska, 1972). Reduction in size of fronds as well as in numbers produced was observed for *Salvinia* in zinc solutions of 1 ppm and higher (Burton, unpublished data). In both these studies, high metal accumulation accompanied the growth effects, in contrast to silver which had an inhibiting effect on growth of *Salvinia* at 0·5 ppm but very little accumulation occurred (Hutchinson and Czyrska, 1975).

4. TRACE METAL TOLERANCE IN LICHENS AND BRYOPHYTES

Different species vary in their sensitivity or tolerance to metals. Photosynthetic ^{14}C fixation by *Stereocaulon paschale* was reduced by several metals to a lesser extent than that of *Umbilicaria muhlenbergii* (Puckett, 1976). The phosphatase activity of *Cladina rangiferina* was lowered by vanadium but was not affected in *C. arbuscula* and *U. mammulata* (LeSueur and Puckett, 1980). Differences in sensitivity were also indicated by the investigations of Biebl (1947, 1950*a*,*b*) referred to previously, and of Url (1956) whose data indicated species of 'copper mosses', particularly *Mielichhoferia elongata*, showed the highest plasmatic resistance to copper of the species tested. Field observations also reflect the different sensitivities of bryophytes and lichens to metals (Lambinon *et al.*, 1964; Folkeson, 1979*a*). The ability to tolerate high concentrations of lead and zinc in mine streams has been demonstrated for some species of aquatic bryophytes (McLean and Jones, 1975; Burton and Peterson, 1979*a*). The species diversity of the polluted streams was considerably less than non-contaminated streams in the vicinity. The liverwort *Scapania undulata* was recorded in a wide range of lead and zinc levels and was frequently the only species present. McLean and Jones (1975) recorded that *Fontinalis squamosa*, transplanted into a more polluted stream showed signs of death and decay after 18 weeks, and a considerably elevated metal content. *Solenostoma crenulatum* is another species able to tolerate high concentrations of lead and zinc, being the only species present in a zinc-rich adit effluent (Burton and Peterson, 1979*a*). Mosses have also been recorded in rivers containing thallium (in addition to other metals) in a mining area (Zitko *et al.*, 1975).

In the context of discussing the ability of lichens to tolerate large metal accumulations, Lange and Ziegler (1963) mention several factors which may aid in interpreting the different lichen responses. These factors include: (1) Inherent cytoplasmic resistance to metallic ions, (2) immobilization of the ions within the cytoplasm by means of chelators and other metal-binding substances, and (3) active and passive transport of the ions to regions external to the plasma and cell wall, resulting in localization of insoluble metallic salts on the surface of the lichens. An example of this latter factor may be the observation of Shimwell and Laurie (1972) of a crystalline deposit on carpets of the moss *Dicranella varia*. The secretion contained lead and zinc in concentrations four and six-fold in excess of those in the moss. Insight into the metal tolerance of lichens can be gained from the studies of sulphur dioxide phytotoxicity (Wirth and Turk, 1975). Lichen thalli were exposed to sulphur dioxide in air or were submerged in aqueous sulphur dioxide solutions. The measure of tolerance was based on the rate of carbon dioxide exchange; and arrangement of the species, according to their response to the gaseous treatment, differed largely from the order for the experiments in the aqueous medium. Wirth and Turk (1975) discussed these differing sensitivities in terms of avoidance (gaseous exposure) or protoplasmic resistance (aqueous exposure). Avoidance is achieved by reducing the amount of pollutant the plant accumulates. For lichens, the wettability and retentive characteristics of the thallus are important in this respect. Surface area, growth form and physical barriers, e.g. cortex structure are other considerations. In mosses, growth form has been suggested to affect the extent of metal uptake (Huckabee, 1973; McLean and Jones, 1975). Protoplasmic resistance is dependent on such features as the maturity, vitality, permeability, and buffering capacity of the cells. The comparisons of lichen response to metals described above can only be considered in terms of protoplasmic resistance. With respect to avoidance, considerable variation in element composition of different species within one habitat have been recorded in field surveys (and in laboratory uptake studies) indicating selective exclusion or absorption of metals. Comparisons of several lichen species showed consistent differences in their element contents. Species of the genus *Cladina* could be discriminated on the basis of their sulphur content (Puckett and Finegan, 1977) or their titanium, iron, scandium or vanadium contents (Puckett and Finegan, 1980). Such differences are clear enough to allow Folkeson

(1979*b*) to obtain interspecies calibration factors allowing estimates of element concentrations for a particular species of lichen or moss to be made on the basis of analyses of other species.

Differences in tolerance between populations of the same lichen species have not been demonstrated, but genetic differences in lead tolerance between populations of the liverwort *Marchantia polymorpha* have been described by Briggs (1972). Gemmae from an urban population in Glasgow developed on agar both in the presence and absence of lead, whereas growth was significantly reduced in the presence of lead when gemmae from a site outside the city were tested. This suggested the latter population would be more sensitive to lead in the habitat. A recent report suggests that copper-tolerant ecotypes of *Solenostoma crenulatum* are present at a copper mine (Brown and House, 1978). In laboratory tests in which copper was supplied, the copper-mine population showed lower potassium loss, no effect on ^{14}C fixation, and apparent control of copper uptake in contrast to a lead-mine population. For lichens, the only examples of differing tolerances are between different species or genera (Nash, 1975; Puckett, 1976; Sheridan, 1979). For higher plants, it has been shown that the greater the degree of metal tolerance, the greater is the metal binding capacity of the cell wall (Turner and Marshall, 1971, 1972), but it has not been determined whether the ability to bind potentially toxic metals extracellularly confers any tolerance on lichens. Although a comparison of metal uptake by *Umbilicaria muhlenbergii* and *Stereocaulon paschale* shows no differences in capacity (Puckett *et al.*, 1973; Puckett, 1974) there are differences in their tolerance to the metals as reflected by subsequent photosynthetic ^{14}C fixation. Zinc-tolerant higher plants take up more zinc than do non-tolerant plants and the zinc taken up is mainly bound to cell-wall sites. The initial uptake and binding of zinc by cell-wall fractions of the grass *Agrostis tenuis* has many features in common to that described above for lichens and bryophytes.

The high capacity for metal accumulation, with often no visible effects on bryophytes, has led to studies on localization. Brown and Bates (1972) concluded that the majority of the lead in *Grimmia doniana* collected from a lead mine, was ionically bound to extracellular sites. Zinc in aquatic bryophytes has also been reported as chiefly cell-wall bound, although a small soluble component was also identified (Burton and Peterson, 1979*b*), as also indicated for copper in *Solenostoma crenulatum* (Brown and House, 1978). Electron microscopy has revealed the presence of lead precipitates in the cell wall of *Hylocomium splendens* and also within the cytoplasm of *Rhytidiadelphus squarrosus* (Skaar *et al.*, 1973). Mosses from an urban polluted area which were watered with high concentrations of lead (1000 and 10 000 ppm) showed evidence of pinocytotic uptake of lead electron-dense particles enclosed in vesicles and in plasma membrane invaginations into the cytoplasm of *R. squarrosus*. In the polluted material, electron-dense inclusions were observed in the cell nuclei of the older leaves, and vacuoles were small. Nuclear inclusions were not seen in experimentally polluted material. Electron-dense lead particles have also been observed in other organelles—chloroplasts, mitochondria and microbodies—and also in plasmodesmata between cells (Ophus and Gullvag, 1974). The presence of lead in nuclei and chloroplasts was confirmed by X-ray microanalysis. Immobilization of lead in the cell wall and also in membrane-bound form would reduce the toxic effects on the cells.

Many species in the plant groups considered in this chapter are able to accumulate very high concentrations of trace elements as documented by the extensive literature on metal contents. It is clear, however, that there is insufficient information relating the relevance of these accumulations to the plant. For example, 'effects' studies have been conducted mainly in the laboratory with little relation to the field situation. Also, there is a definite lack of information on the interaction of trace metals with ferns and horsetails.

REFERENCES

Alexander, V. (1974). A synthesis of the IBP Tundra Biome circumpolar study of nitrogen fixation. In: *Soil Organisms and Decomposition in Tundra* (A. J. Holding, O. W. Heal, S. F. MacLean and P. W. Flanagan (eds)), Tundra Biome Steering Committee, Stockholm, pp. 109–21.

Barrie, L. A. (1978). The concentration and deposition of sulphur compounds and metals around the GCOS oil extraction plant during June, 1977. AOSERP Sub-project ME1.5.3, Atmospheric Environment Service, Environment Canada, 4905 Dufferin St., Downsview, Ontario, Canada.

Barrie, L. A. and J. Kovalick (1978). A winter-time investigation of the deposition of pollutants around an isolated power plant in northern Alberta. Internal Report ARQT-4-78. Atmospheric Dispersion Division, Atmospheric Environment Service, Environment Canada, 4905 Dufferin St., Downsview, Ontario, Canada.

Barry, E. F., M. T. Rei and H. H. Reynolds (1975). Determination of nickel in the atmosphere of eastern Massachusetts. *Environ. Lett.* **8**: 381–5.

Bauer, E. and K. Kreeb (1974). Flechtenkartierung und Enzymaktivitat als Indikatoren der Luftverunreinigung in Esslingen. Verh. Ges. Okologie, Saarbrucken 1973, pp. 273–81.

Beckett, P. J. and S. J. Wainwright (1980). Field studies of zinc toxicity in *Parmelia saxatilis* (L.) Ach. *New Phytol.* (in press).

Beschel, R. E. (1959). The copper swamp in the Aboushagan Woods, North Sackville, New Brunswick. Summary for the Maritime Excursion after the 9th International Botanical Congress, Montreal, 7 September 1959 (mimeographed).

Biebl, R. (1947). Uber die gegensatzliche Wirkung der Spurenelemente Zink und Bor auf die Blattzellen von *Mnium rostratum*. *Ost. Bot. Z.* **94**: 61–73.

Biebl, R. (1950*a*). Vergleichende chemische Resistenzstudien an pflanzlichen Plasmen. *Protoplasma*. **39**: 1–13.

Biebl, R. (1950*b*). Uber die Resistenz pflanzlicher Plasmen gegen Vanadium. *Protoplasma*. **39**: 251–9.

Biebl, R. and W. Rossi-Pillhofer (1955). Die Anderung der chemischen Resistenz pflanzlicher Plasmen mit dem Entwicklungszustand. *Protoplasma*. **44**: 113–35.

Bliss, L. C. (1962). Adaptations of arctic and alpine plants to environmental conditions. *Arctic*. **15**: 117–44.

Boissière, M. C. (1973). Activité phosphatasique neutre chez le phycobionte de *Peltigera canina* comparée à celle d'un *Nostoc* libre. *C.R. Acad. Sci. Paris*. **277** (Série D): 1649–51.

Briggs, D. (1972). Population differentiation in *Marchantia polymorpha* L. in various lead pollution levels. *Nature*. **238**: 166–7.

Brodo, I. M. (1973). Substrate ecology. In: *The Lichens* (V. Ahmadjian and M. E. Hale (eds)), Academic Press, New York, pp. 401–41.

Brown, D. H. (1976). Mineral uptake by lichens. In: *Lichenology: Progress and Problems. System. Assoc. Special Vol. 8* (D. H. Brown, D. L. Hawksworth and R. H. Bailey (eds)), Academic Press, London, pp. 419–39.

Brown, D. H. and J. W. Bates (1972). Uptake of lead by two populations of *Grimmia doniana*. *J. Bryol.* **7**: 187–93.

Brown, D. H. and D. R. Slingsby (1972). The cellular location of lead and potassium in the lichen *Cladonia rangiformis* (L.) Hoffm. *New Phytol.* **71**: 297–305.

Brown, D. H. and K. L. House (1978). Evidence of a copper-tolerant ecotype of the hepatic *Solenostoma crenulatum*. *Ann. Bot.* **42**: 1383–92.

Burkitt, A., P. Lester and G. Nickless (1972). Distribution of heavy metals in the vicinity of an industrial complex. *Nature*. **238**: 327–8.

Burton, M. A. S. and P. J. Peterson (1979*a*). Metal accumulation by aquatic bryophytes from polluted mine streams. *Environ. Pollut.* **19**: 39–46.

Burton, M. A. S. and P. J. Peterson (1979*b*). Studies on zinc localization in aquatic bryophytes. *Bryologist*. **82**: 594–8.

Burton, M. A. S., P. LeSueur and K. J. Puckett (1981). Copper, nickel and thallium uptake by the lichen *Cladina rangiferina*. *Can. J. Bot.* **59**: 91–100.

Cannon, H. L., H. T. Shacklette and H. Bastron (1968). Metal absorption by *Equisetum* (Horsetail), US Geological Survey Bulletin, No. 1278-A.

Chimiklis, P. E. and R. L. Heath (1975). Ozone-induced loss of intracellular potassium ion from *Chlorella sorokiniana*. *Plant Physiol.* **56**: 723–7.

Connor, J. J. (1979). Geochemistry of ohia and soil lichen, Puhimau thermal area, Hawaii. *Sci. Total Env.* **12**: 241–50.

Coombes, A. J. and N. W. Lepp (1974). The effect of Cu and Zn on the growth of *Marchantia polymorpha* and *Funaria hygrometrica*. *Bryologist*. **77**: 447–52.

Coppins, B. J. (1978). A glimpse of the past and present flora of Edinburgh. *Trans. Proc. Bot. Soc. Edinb.* **42**: 19–35.

Crespo, A., E. Manrique, E. Barreno and E. Serina (1977). Valaracion de la contaminacion atmosferica del area urbana de Madrid mediante bioindicadores (liquenes epifitos). *An. Inst. Bot. A. J. Cavanillo.* **34**: 71–94.

Doyle, P., W. K. Fletcher and V. C. Brink (1973). Trace element content of soils and plants from the Selwyn mountains, Yukon and Northwest Territories. *Can. J. Bot.* **51**: 421–7.

Folkeson, L. (1979*a*). Changes in the cover of mosses and lichens in coniferous woodland polluted with copper and zinc. In: *The Use of Ecological Variables in Environmental Monitoring, Proceedings of the First Nordic Oikos Conference, 1978* (H. Hytteborn (ed)), Uppsala, Sweden, pp. 59–61.

Folkeson, L. (1979*b*). Interspecies calibration of heavy-metal concentrations in nine mosses and lichens: Applicability to deposition measurements. *Water Air Soil Poll.* **11**: 253–60.

Foy, C. D., R. L. Chaney and M. C. White (1978). The physiology of metal toxicity in plants. *Ann. Rev. Plant Physiol.* **29**: 511–66.

Garty, J., M. Galun and M. Kessel (1979). Localization of heavy metals and other elements accumulated in the lichen thallus. *New Phytol.* **82**: 159–68.

Goodman, G. T. and T. M. Roberts (1971). Plants and soils as indicators of metals in the air. *Nature.* **231**: 287–92.

Gordon, G. E., W. H. Zoller and E. S. Gladney (1974). Abnormally enriched trace elements in the atmosphere. In: *Trace Substances in Environmental Health* (D. D. Hemphill (eds)), University of Columbia, Missouri, pp. 161–6.

Groet, S. S. (1976). Regional and local variations in heavy metal concentrations of bryophytes in the northeastern United States. *Oikos.* **27**: 445–56.

Haag, R. W. (1974). Nutrient limitations to plant production in two tundra communities. *Can. J. Bot.* **52**: 106–16.

Handley, R. and R. Overstreet (1968). Uptake of carrier-free ^{137}Cs by *Ramalina reticulata. Plant Physiol.* **43**: 1401–5.

Harris, T. M. (1946). Zinc poisoning of wild plants from wire netting. *New Phytol.* **45**: 50–5.

Henriksson, L. E. and E. J. DaSilva (1978). Effects of some inorganic elements on nitrogen fixation in blue–green algae and some ecological aspects of pollution. *Zeitschr. Allg. Mikrobiol.* **18**: 487–94.

Hocking, D. and R. A. Blauel (1977). Progressive heavy metal accumulation associated with forest decline near the nickel smelter at Thompson, Manitoba. Environment Canada, Can. For. Serv., North For. Res. Cent. Inf. Rep. NOR-X-169. 20 pp.

Hocking, D. and M. B. Hocking (1977). Equilibrium solubility of trace atmospheric sulphur dioxide in water and its bearing on air pollution injury to plants. *Environ. Pollut.* **13**: 57–63.

Hocking, D., P. Kuchar, J. A. Plambeck and R. A. Smith (1978). The impact of gold smelter emissions on vegetation and soils of a sub-Arctic forest–tundra transition ecosystem. *J. Air. Poll. Contr. Assoc.* **28**: 133–7.

Holt, A. S. (1958). The phase test intermediate and the allomerisation of chlorophyll *a*. *Can. J. Biochem. Physiol.* **36**: 439–56.

Huckabee, J. W. (1973). Mosses: Sensitive indicators of airborne mercury pollution. *Atmos. Env.* **7**: 749–54.

Hunter, J. G. (1953). The composition of bracken: Some major- and trace-element constituents. *J. Sci. Food Agric.* **4**: 10–20.

Huss-Danell, K. (1977). Nitrogen fixation by *Stereocaulon paschale* under field conditions. *Can. J. Bot.* **55**: 585–92.

Hutchinson, T. C. and H. Czyrska (1972). Cadmium and zinc toxicity and synergism to floating aquatic plants. *Water Poll. Research Canada.* **7**: 59–75.

Hutchinson, T. C. and H. Czyrska (1975). Heavy metal toxicity and synergism to floating aquatic weeds. *Verh. Internat. Verein. Limnol.* **19**: 2102–11.

Jaakkola, T., H. Puumala and J. K. Miettinen (1966). Microelement levels in environmental samples in Finland. In: *Radioecological Concentration Processes. Proceedings of International Symposium* (B. Aberg and F. P. Hungate (eds)), Pergamon Press, Oxford, pp. 341–50.

James, P. (1973). The effect of air pollutants other than hydrogen fluoride and sulphur dioxide on lichens. In: *Air Pollution and Lichens* (B. W. Ferry, M. S. Baddeley and D. L. Hawksworth (eds)), Athlone Press, London, pp. 143–75.

Johannessen, M. and A. Henriksen (1978). Chemistry of snow meltwater. Changes in concentration during melting. *Water Resources Research.* **4**: 615–19.

Johnsen, I. (1975). I. The uptake of lead, cadmium and vanadium by lichens and bryophytes. II. The effects of SO_2 fumigation on lichens and bryophytes and its interaction with lead uptake from water solutions. Research Report JRC-EURATOM, Ispra, Italy.

Johnston, L. G. and W. F. Watson (1956). The allomerization of chlorophyll. *J. Chem. Soc.* pp. 1203–12.

Laaksovirta, K. and H. Olkkonen (1977). Epiphytic lichen vegetation and element contents of *Hypogymnia physodes* and pine needles examined as indicators of air pollution at Kokkola, W. Finland. *Ann. Bot. Fennici.* **14**: 112–30.

Lambinon, J., A. Maquinay and J. L. Ramaut (1964). La teneur en zinc de quelques lichens des terrains calaminaires belges. *Bull. Jardn. Bot. Etat. Brux.* **34**: 273–82.

Lane, I. and K. J. Puckett (1979). Responses of the phosphatase activity of the lichen *Cladina rangiferina* to various environmental factors including metals. *Can. J. Bot.* **57**, 1534–40.

Lange, O. L. and H. Ziegler (1963). Der Schwermetallgehalt von Flechten aus dem *Acarosporetum sinopicae* auf Erzschlackenhalden des Harzes. *Mitt. flor.-soz. Arb. Gemein.* **10**: 156–83.

Lawrey, J. D. and M. E. Hale (1979). Lichen growth responses to stress induced by automobile exhaust pollution. *Science.* **204**: 423–4.

LeBlanc, F. and J. De Sloover (1970). Relation between industrialization and the distribution and growth of epiphytic lichens and mosses in Montreal. *Can. J. Bot.* **48**: 1485–96.

LeBlanc, F., G. Robitaille and D. N. Rao (1974). Biological response of lichens and bryophytes to environmental pollution in the Murdochville copper mine area, Quebec. *J. Hattori Bot. Lab.* **38**: 405–33.

Lee, J., R. R. Brooks, R. D. Reeves and T. Jaffré (1977). Chromium-accumulating bryophyte from New Caledonia. *Bryologist.* **80**: 203–5.

Lepp, N. W. and M. J. Roberts (1977). Some effects of cadmium on growth of bryophytes. *Bryologist.* **80**: 533–6.

Lerond, M. (1978). Courbes d'isopollution de la région de Rouen obtenues par l'observation des lichens epiphytes. *Bull. Soc. Linn. Normandie.* **106**: 73–84.

LeSueur, P. and K. J. Puckett (1980). Effect of vanadium on the phosphatase activity of lichens. *Can. J. Bot.* **58**: 502–4.

Lounamaa, K. J. (1956). Trace elements in plants growing wild on different rocks in Finland. A semi-quantitative survey. *Annls. Bot. Soc. Zool. Bot. Fenn. 'Vanamo'*. **29**: 1–196.

Malaisse, F., J. Gregoire, R. S. Morrison, R. R. Brookes and R. D. Reeves (1979). Copper and cobalt in vegetation of Fungurume, Shaba Province, Zaire. *Oikos.* **33**: 472–8.

Massé, L. (1964). Réchérches phytosociologiques et écologiques sur les lichens des schistes rouges Cambriens des environs de Rennes. *Vegetatio.* **12**: 103–216.

Maquinay, A., I. M. Lamb, J. Lambinon and J. L. Ramaut (1961). Dosage du zinc chez un lichen calaminaire belge: *Stereocaulon nanodes* Tuck. f. *tyroliense* (Nyl.) M. Lamb. *Physiol. Plant.* **14**: 284–9.

McLean, R. O. and A. K. Jones (1975). Studies of tolerance to heavy metals in the flora of the rivers Ystwyth and Clarach, Wales. *Freshwat. Biol.* **5**: 431–44.

Nash, T. H. (1975). Influence of effluents from a zinc factory on lichens. *Ecol. Monog.* **45**: 183–98.

Nieboer, E., P. Lavoie, R. L. P. Sasseville, K. J. Puckett and D. H. S. Richardson (1976*a*). Cation-exchange equilibrium and mass balance in the lichen *Umbilicaria muhlenbergii*. *Can. J. Bot.* **54**: 720–3.

Nieboer, E., K. J. Puckett and B. Grace (1976*b*). The uptake of nickel by *Umbilicaria muhlenbergii*: A physicochemical process. *Can. J. Bot.* **54**: 724–33.

Nieboer, E., D. H. S. Richardson and F. D. Tomassini (1978). Mineral uptake and release by lichens: An overview. *Bryologist.* **81**: 226–46.

Nieboer, E., D. H. S. Richardson, P. Lavoie and D. Padovan (1979). The role of metal-ion binding in modifying the toxic effects of sulphur dioxide on the lichen *Umbilicaria muhlenbergii*. I. Potassium efflux studies. *New Phytol.* **82**: 621–32.

Noguchi, A. and H. Furuta (1956). Germination of spores and regeneration of leaves of *Merceya ligulata* and *M. gedeana*. *J. Hattori Bot. Lab.* **17**: 32–44.

Ophus, E. M. and B. M. Gullvag (1974). Localization of lead within leaf cells of *Rhytidiadelphus squarrosus* (Hedw.) Warnst. by means of transmission electron microscopy and X-ray microanalysis. *Cytobios.* **10**: 45–8.

Pakarinen, P. and K. Tolonen (1976). Regional survey of heavy metals in peat mosses (*Sphagnum*). *Ambio.* **5**: 38–40.

Persson, H. (1956). Studies in 'copper mosses'. *J. Hattori Bot. Lab.* **17**: 1–18.

Peterson, P. J., M. A. S. Burton, M. Gregson, S. M. Nye and E. K. Porter (1976). Tin in plants and surface waters in Malaysian ecosystems. In: *Trace Substances in Environmental Health*—X (D. D. Hemphill (ed)), University of Columbia, Missouri, pp. 123–32.

Pickering, D. C. and I. L. Puia (1969). Mechanism for the uptake of zinc by *Fontinalis antipyretica*. *Physiol. Plant.* **22**: 653–61.

Pilegaard, K. (1978). Airborne metals and SO_2 monitored by epiphytic lichens in an industrial area. *Environ. Pollut.* **17**: 81–92.

Poelt, J. and S. Huneck (1968). *Lecanora vinetorum* nova spec., ihre Vergesellschaftung, ihre Okologie und ihre Chemie. *Osterr. Bot. Z.* **115**: 411–22.

Puckett, K. J. (1974). The ecology and physiology of lichens with respect to atmospheric pollution. Ph.D. Thesis, University of London, UK. 313 pp.

Puckett, K. J. (1976). The effect of heavy metals on some aspects of lichen physiology. *Can. J. Bot.* **54**: 2695–703.

Puckett, K. J. (1978). Element levels in lichens from the Northwest Territories. Report ARQA-56-78. Atmospheric Environment Service, Environment Canada.

Puckett, K. J. and E. J. Finegan (1977). The use of multivariate techniques in the analysis of metal levels found in lichens. Report ARQA-52-77. Atmospheric Environment Service, Environment Canada.

Puckett, K. J. and E. J. Finegan (1980). An analysis of the element content of lichens from the Northwest Territories, Canada. *Can. J. Bot.* **58**: 2073–89.

Puckett, K. J., E. Nieboer, M. J. Gorzynski and D. H. S. Richardson (1973). The uptake of metal ions by lichens: A modified ion-exchange process. *New Phytol.* **72**: 329–42.

Puckett, K. J., F. D. Tomassini, E. Nieboer and D. H. S. Richardson (1977). Potassium efflux by lichen thalli following exposure to aqueous sulphur dioxide. *New Phytol.* **79**: 133–45.

Pyatt, F. B. (1976). Lichen ecology of metal spoil tips: Effects of metal ions on ascospore viability. *Bryologist.* **79**: 172–9.

Rasmussen, L. (1977). Epiphytic bryophytes as indicators of the changes in the background levels of airborne metals from 1951–75. *Environ. Pollut.* **14**: 37–45.

Rastorfer, J. R. (1974). Element contents of three Alaskan–Arctic mosses. *Ohio J. Science.* **74**: 55–9.

Rao, D. N., G. Robitaille and F. LeBlanc (1977). Influence of heavy metal pollution on lichens and bryophytes. *J. Hattori Bot. Lab.* **42**: 213–39.

Ray, S. and W. White (1976). Selected aquatic plants as indicator species for heavy metal pollution. *J. Environ. Sci. Health.* **A11**(12): 717–25.

Ray, S. N. and W. J. White (1979). *Equisetum arvense*—An aquatic vascular plant as a biological monitor for heavy metal pollution. *Chemosphere.* **3**: 125–8.

Rejment-Grochowska, I. (1976). Concentration of heavy metals, lead, iron, manganese, zinc and copper in mosses. *J. Hattori Bot. Lab.* **41**: 225–30.

Richardson, D. H. S., E. Nieboer, P. Lavoie and D. Padovan (1979). The role of metal-ion binding in modifying the toxic effects of sulphur dioxide on the lichen *Umbilicaria muhlenbergii.* II. ^{14}C fixation studies. *New Phytol.* **82**: 633–43.

Ritter-Studnicka, H. and O. Klement (1968). Uber Flechtenarten und deren Gesellschaften auf Serpentin in Bosnien. *Osterr. Bot. Z.* **115**: 93–9.

Roberts, T. M. and G. T. Goodman (1974). The persistence of heavy metals in soils and natural vegetation following closure of a smelter. In: *Trace Substances in Environmental Health*—VIII (D. D. Hemphill (ed)), University of Columbia, Missouri, pp. 117–25.

Robinson, W. O., H. W. Lakin and L. E. Reichen (1947). The zinc content of plants on the Friedensville zinc slime ponds in relation to biogeochemical prospecting. *Econ. Geology.* **42**: 572–82.

Ross, I. S. (1975). Some effects of heavy metals on fungal cells. *Trans. Br. Mycol. Soc.* **64**: 175–93.

Ruhling, A. and G. Tyler (1970). Sorption and retention of heavy metals in the woodland moss *Hylocomium splendens* (Hedw.) Br. et Sch. *Oikos.* **21**: 92–7.

Ruhling, A. and G. Tyler (1971). Regional differences in the deposition of heavy metals over Scandinavia. *J. Appl. Ecol.* **8**: 497–507.

Salmon, L., D. H. F. Atkins, R. Fisher, C. Healy and D. V. Law (1978). Retrospective trend analysis of the content of UK air particulate matter 1974–1975. *Sci. Total Env.* **9**: 161–200.

Schatz, A. (1955). Speculations on the ecology and photosynthesis of the 'copper mosses'. *Bryologist.* **58**: 113–19.

Schmid, M. L. and K. Kreeb (1975). Enzymatische Indikation gasgeschaedigter Flechten. *Angew. Botanik.* **49**: 141–54.

Schutte, J. A. (1976). The accumulation of chromium by two lichen species of the lichen genus Parmelia and the subsequent effects. Ph.D. Thesis, Ohio State University, Columbus, Ohio.

Schutte, J. A. (1977). Chromium in two corticolous lichens from Ohio and West Virginia. *Bryologist.* **77**: 279–84.

Scotter, G. W. and J. E. Miltimore (1973). Mineral content of forage plants from the reindeer preserve, Northwest Territories. *Can. J. Plant Sci.* **53**: 263–8.

Seaward, M. R. D. (1974). Some observations on heavy metal toxicity and tolerance in lichens. *Lichenologist.* **6**: 158–64.

Seeley, G. R. (1966). The structure and chemistry of function groups. In: *The Chlorophylls* (C. P. Vernon and G. R. Seeley (eds)), Academic Press, New York, pp. 67–109.

Sergy, G. (1978). Environmental distribution of cadmium in the Prairie Provinces and Northwest Territories. Economic and Technical Review Report EPS 3-NW-78-2, Environmental Protection Service, Fisheries and Environment Canada.

Shacklette, H. T. (1961). Substrate relationships of some bryophyte communities on Latouche Island, Alaska. *Bryologist.* **64**: 1–16.

Shacklette, H. T. (1965). Element content of bryophytes. US Geological Survey Bulletin, No. 1198-D.

Shacklette, H. T. (1967). Copper mosses as indicators of metal concentrations. US Geological Survey Bulletin, No. 1198-G.

Sherbin, I. G. (1979). Mercury in the Canadian environment. Economic and Technical Review Report EPS 3-EC-79-6. Environmental Protection Service, Environment Canada.

Sheridan, R. P. (1979). Impact of emissions from coal-fired electricity generating facilities on N_2-fixing lichens. *Bryologist.* **82**: 54–8.

Shimwell, D. W. and A. E. Laurie (1972). Lead and zinc contaminations of vegetation in the southern Pennines. *Environ. Pollut.* **3**: 291–301.

Simola, L. K. (1976). The effect of some heavy metals, and fluoride and arsenate ions on the growth and fine structure of *Sphagnum fimbriatum* in aseptic culture. In: *Proceedings of the Kuopio Meeting on Plant Damages Caused by Air Pollution* (L. Karenlampi (ed)), Kuopio, Finland, pp. 148–54.

Simola, L. K. (1977). The effect of lead, cadmium, arsenate, and fluoride ions on the growth and fine structure of *Sphagnum nemoreum* in aseptic culture. *Can. J. Bot.* **55**: 426–35.

Skaar, H., E. Ophus and B. M. Gullvag (1973). Lead accumulation within nuclei of moss leaf cells. *Nature.* **241**: 215–16.

Solberg, Y. and A. R. Selmer-Olsen (1978). Studies on the chemistry of lichens and mosses. XVII. Mercury content of several lichen and moss species collected in Norway. *Bryologist*. **81**: 144–9.

Steinnes, E. (1977). Atmospheric deposition of trace elements in Norway studied by means of moss analysis. Institute for Atomic Energy, Kjeller Research Establishment, Kjeller, Norway. 13 pp.

Syers, J. K., A. C. Birnie and B. D. Mitchell (1967). The calcium oxalate content of some lichens growing on limestone. *Lichenologist*. **3**: 409–14.

Tomassini, F. D. (1976). The measurement of photosynthetic ^{14}C fixation rates and potassium efflux to assess the sensitivity of lichens to sulphur dioxide and the adaptation of X-ray fluorescence to determine the elemental content in lichens. M.Sc. Thesis, Laurentian University, Sudbury, Ontario.

Tomassini, F. D., K. J. Puckett, E. Nieboer, D. H. S. Richardson and B. Grace (1976). Determination of copper, iron, nickel, and sulphur by X-ray fluorescence in lichens from the Mackenzie Valley, Northwest Territories and the Sudbury district, Ontario. *Can. J. Bot.* **54**: 1591–1603.

Tuominen, Y. (1967). Studies on the strontium uptake of the *Cladonia alpestris* thallus. *Ann. Bot. Fenn.* **4**: 1–27.

Turner, R. G. and C. Marshall (1971). The accumulation of ^{65}Zn by root homogenates of zinc tolerance and non-tolerant clones of *Agrostis tenuis* Sibth. *New Phytol.* **70**: 539–45.

Turner, R. G. and C. Marshall (1972). The accumulation of zinc by subcellular fractions of roots of *Agrostis tenuis* Sibth., in relation to zinc tolerance. *New Phytol.* **71**: 671–6.

Url, W. (1956). Über Schwermetall-, zumal Kupferresistenz einiger Moose. *Protoplasma*, **46**: 768–93.

Wainwright, S. J. and P. J. Beckett (1975). Kinetic studies on the binding of zinc by the lichen *Usnea florida* (L.) Web. *New Phytol.* **75**: 91–8.

Wallin, T. (1976). Deposition of airborne mercury from six Swedish chlor-alkali plants surveyed by moss analysis. *Environ. Pollut.* **10**: 101–14.

Ward, N. I., R. R. Brooks and E. Roberts (1977). Heavy metals in some New Zealand bryophytes. *Bryologist*. **80**: 304–12.

Wetmore, C. M. (1970). The lichen family *Heppiaceae* in North America. *Ann. Mo. Bot. Gard.* **57**: 158–209.

Whitehead, N. E. and R. R. Brooks (1969). Aquatic bryophytes as indicators of uranium mineralization. *Bryologist*. **72**: 501–7.

Wirth, V. (1972). Die Silikatflechten-Gemeinschaften in ausseralpinen Zentraleuropa. *Dissert. Bot.* **17**: 1–306.

Wirth, V. and R. Turk (1975). Zur SO_2-Resistenz von Flechten verschiedener Wuchsform. *Flora*. **164**: 133–43.

Zitko, V., W. V. Carson and W. G. Carson (1975). Thallium: Occurrence in the environment and toxicity to fish. *Bull. Environ. Contam. Toxicol.* **13**: 23–30.

CHAPTER 8

Mechanism of Metal Tolerance in Higher Plants

D. A. THURMAN
Department of Botany, University of Liverpool, UK

1. INTRODUCTION

A number of higher plants have evolved populations with the ability to grow and flourish in soils containing elevated levels of trace metals. These populations are said to be metal tolerant. This character has been found amongst both wild and crop species. Wild species often show differentiation into ecotypes which can vary in their degree of tolerance; some cultivars of various crop species also show variation in their degree of tolerances.

Over the past several years, studies on heavy metal tolerance have become more broadly based, the initial studies on methods used for measuring tolerance and investigations on ecological aspects of the subject have been complemented more recently by genetical, physiological and biochemical approaches. This has resulted in the accumulation of information along a broader front, rather than a marked advancement in any one particular area. This applies to studies on the mechanisms of metal tolerance operating in edaphic ecotypes which are the main concern of this chapter. At this time we do not have a clear-cut picture of what types of mechanism are operating in these ecotypes which are tolerant to various heavy metals. What follows is a discussion of some of the mechanisms which are thought to operate in these types of plants.

2. EXCLUSION OF METAL FROM THE PLANT

For many heavy metal tolerant grasses, exclusion of metal from the roots and/or shoots does not appear to form part of the tolerance mechanism.

Zinc-tolerant clones of *Agrostis tenuis*, *A. stolonifera* (Peterson, 1969; Wainwright and Woolhouse, 1975) and *Deschampsia caespitosa* (Brookes *et al.*, 1980) all accumulate as much zinc in their roots and shoots as do non-tolerant clones; these clones show no evidence of metal exclusion. There are indications that this is not always the case in other species. Baker (1978*a*), working with a number of different populations of *Silene maritima* from zinc-contaminated soils, observed that all populations from these sites, when grown in culture for 12 weeks in 5 ppm zinc, exclude zinc from their shoots; some of the mine populations also showed a reduced overall uptake.

The accumulation of metal in the roots by some tolerant ecotypes, with restriction of movement of metal to the shoots, has prompted the suggestion that this plays a part in the tolerance mechanism. Wu *et al.* (1975) showed that copper-tolerant material of *A. stolonifera* accumulated copper in the roots and so restricted its entry into the shoots. This was in marked contrast to the non-tolerant plants, where copper passed into the shoot. From the above discussion, restruction of movement of metal to the shoot cannot be a universal mechanism for all species and metals. Furthermore, a number of non-tolerant species are known to store metals in their roots when growing under 'luxury levels' of the element. West (1979) showed that *Pinus radiata* can store copper in its roots when growing in a plentiful supply of the element and subsequently release it to the shoot when faced with restricted supplies.

How one heavy metal may affect the uptake of another by a tolerant plant has not been investigated in any great detail. Wu and Antonovics (1975) have examined the effect of copper upon zinc uptake and vice versa by two clones of *A. stolonifera*, one non-tolerant and another co-tolerant to zinc and copper. These authors made the claim, from what appears to be variable and limited data, that copper and zinc uptake by these clones takes place independently. This observation requires more extensive investigation over a much wider range of metal concentrations before precise statements can be made about whether or not one metal can influence the uptake of another by tolerant plants. In studies of this type it would be informative if indications were given upon the viability of the material analysed, especially for roots, which can accumulate high levels of metal.

3. THE ROLE OF THE CELL WALL

The hypothesis that cell walls play an important role in metal tolerance was first suggested by Turner (1967, 1969) using a copper- and a zinc-tolerant clone of *A. tenuis*. Analysis of subcellular fractions of roots of this plant showed that cell wall preparations of tolerant plants had a superior capacity for binding metals than cell walls of non-tolerant plants. It was further suggested (Turner and Marshall, 1972) that this could exert a protective action by preventing zinc reaching sensitive sites in the roots. Turner and Marshall (1971), again after an examination of the binding of metals by subcellular components, suggested that the metabolism of tolerant plants could also be adapted to high levels of metal.

The substance(s) responsible for binding zinc in the walls of tolerant plants has never been isolated. Peterson (1969), using ^{65}Zn, found that a pectate extract of cell walls of a zinc-tolerant clone of *A. stolonifera* contained five to six times more zinc than a similar extract of cell walls from a non-tolerant clone of this species. The pectate fraction was isolated by its solubility characteristics and was not characterized further. If, as this work suggests, carboxyl groups are important in metal binding, then zinc-tolerant clones of *A. stolonifera* should also be copper tolerant, which they are not, for copper has a greater affinity for carboxyl groups than zinc (Gund and Wilcox, 1956). Brookes (1979) has measured the root cation-exchange capacity of zinc-tolerant and non-tolerant roots of *D. caespitosa* and *A. odoratum* and showed that they are similar, a result not to be expected if acidic polysaccharides contribute more to wall structure in zinc-tolerant plants. She also measured pK values for cell wall preparations from these two species. These values were all between pH 5·0 and 6·0 and did not differ significantly between tolerant and non-tolerant clones; if acidic polysaccharides formed a major component of the cell walls of tolerant plants, lower values would have been expected.

Baker (1978*b*) again drew attention to the role of pectates in tolerance and was prepared to use this as circumstantial evidence for the involvement of calcium in zinc tolerance. He found from long-term culture experiments with *Silene maritima* that calcium stimulated the uptake of zinc by tolerant plants and interpreted this as a corollary of zinc tolerance in his plants.

Turner and Marshall (1972) discounted cell wall proteins as binding agents for zinc in tolerant *A. tenuis*, though Wainwright and Woolhouse (1975) have questioned this rejection. They pointed out that there is enough nitrogen in the cell wall of this grass to bind zinc and copper to the levels reported by Turner and Marshall. Wyn Jones *et al.* (1971), also working with zinc tolerant *A. tenuis*, reported the isolation of an amino sugar complex which was capable of binding zinc. However, the complex has never been characterized. It is clear from the results discussed here that the exact nature of the binding substance(s), if present in cell walls of tolerant plants, is far from clear. Without its identification the cell wall hypothesis of metal tolerance remains unconfirmed.

4. EVIDENCE FOR THE PASSAGE OF HEAVY METALS INTO THE SYMPLAST

A number of observations suggest that heavy metals do enter the symplast of root and shoot cells of tolerant plants. The uptake studies discussed above show that shoots of metal-tolerant plants often contain as much metal as non-tolerant plants. This fact, together with the observation that bivalent cations must pass through the living cells in the endodermis (Robards *et al.*, 1973) and by so doing, move through the symplast on their way to the shoot, implies that metal must enter the cytoplasm. It is difficult to envisage the metal reaching the shoot from the root entirely by a cell wall pathway and once present in the leaf then residing in the cell wall. Wainwright and Woolhouse (1977) have made the important observation, using root segments of zinc-tolerant, copper-tolerant and non-tolerant clones of *A. tenuis*, that even under conditions in which the capacity of root segments to adsorb metal ions is swamped, the segments of the tolerant plants continued to grow in increasing concentrations of zinc when compared with the non-tolerant strain. These experiments were carried out under conditions which precluded the cell wall forming a significant buffer which might modify the effective concentration of zinc. Unless these root cells are totally impermeable to zinc, which would be inconsistent with what has been stated above, the tolerance mechanism cannot reside solely in the cell wall.

One criticism of trying to determine the distribution of ions by cell fractionation procedures is the possibility of redistribution of ions taking place amongst the separated components, particularly when the fractionations are carried out with polar solvents. To try to overcome this problem, other techniques must be applied and the results compared. Thinking along these lines, Brookes *et al.* (1980) have carried out a compartmental flux analysis using ^{65}Zn with zinc-tolerant and non-tolerant roots of *D. caespitosa*. In this study, the amount of zinc present in the free space of the roots of the two clones was estimated from the initial influx curves. At all levels of zinc tested, the zinc content of the free space was similar in both clones. This observation runs contrary to the results of the cell fractionation studies already described; unfortunately, the two studies were carried out with different species which could possibly account for the differences observed. If this is so, different species may have evolved different mechanisms for the same metal. A broadly based study in which compartmental flux analysis is applied to a number of species tolerant to different metals may resolve this difference. Brookes *et al.* (1980) found that the major difference between their two types of clones of *D. caespitosa* was the ability of the tolerant roots to actively pump zinc into their vacuoles when exposed to external concentrations of zinc up to 1 mM; non-tolerant root cells could not carry out this process above 0·1 mM zinc. It appears in *D. caespitosa* that the majority of the zinc in tolerant roots is present in vacuoles, again implying that the metal must move through the cytoplasm of tolerant cells. If this is so, the question arises as to in what form is the zinc present in the vacuole? The data of Brookes *et al.* (1980) could be interpreted so as to suggest that the majority of zinc is present as ionic zinc, though their results do not exclude complexing of the metal.

5. ORGANIC ACIDS AND METAL TOLERANCE

Complexing of metals by organic acids has been described for a number of metals and acids and is thought to play a role in a number of different physiological processes in plants (Jones, 1961; Grime and Hodgson, 1969; Tiffin, 1972; Lee *et al.*, 1978). Mathys (1977) has suggested a mechanism of zinc tolerance based upon complexing of metals with various organic compounds including organic acids, thus reducing the activity coefficients of the free ions and their toxicity. He suggested that in zinc-tolerant plants, once the metal enters the cell it is complexed with malate. The complex then moves across the tonoplast where the zinc is then complexed in the vacuole with a variety of organic compounds including citrate, oxalate, mustard oils and anthocyanins. To date, this model has little data to support it. Certainly, Ernst (1976) has been able to correlate high levels of malic acid in plants which are either zinc or zinc–copper tolerant. Such plants also showed elevated levels of citrate, but of a lesser order of magnitude compared to malate. Just because high levels of these two acids occur in tolerant plants, there are no *a priori* reasons to suppose that this occurrence is part of the tolerance mechanism. Rankin and Thurman (unpublished data) have attempted to examine this hypothesis further by analysing tolerant and non-tolerant clones of *D. caespitosa* grown under defined cultural conditions in the presence and absence of zinc. Following the citrate and malate levels in the roots over a 10-day period, they found that the amounts of these two acids increased in both clones, though the increase in the levels in tolerant plants was much larger than in the non-tolerant ones. The differences in the levels of citrate between the two clones was particularly striking, being much higher in the tolerant clone grown in the presence of zinc. This contrasts with Ernst's observation that high malate, and not citrate, levels are a feature of zinc-tolerant plants, though it should be pointed out that Ernst's analysis was carried out with plants taken from their natural habitats and not grown in culture solutions. We believe that if high levels of citrate in plants can only be induced by metals to which the plant in question is tolerant, this will provide good evidence for the involvement of citrate in metal tolerance.

If it is accepted that zinc enters the cytoplasm of tolerant cells and citrate does play a role in the tolerance mechanism, the question arises as to how the extra citrate is produced. In most plant cells (except those which contain glyoxysomes) citrate synthesis takes place in mitochondria through the action of citrate synthase from oxaloacetate and acetyl-CoA. Thomas *et al.* (1973) have suggested that in cells which accumulate citrate, the activity of citrate synthase is far above that required to keep the TCA cycle operating during respiration and, without any increase in aconitase, citrate would accumulate. This accumulation would require replenishment of certain cycle intermediates through β-carboxylation in the cytoplasm, additional activity of malate dehydrogenase to produce oxaloacetate, and also of the enzymes producing acetyl-CoA. Malate may also be required to move citrate out of the mitochondria into the cytoplasm on one of the anion transporters which have been described for these organelles. Citrate could then move into the cytoplasm and chelate free zinc ions, reducing its activity coefficient, and so possibly decreasing its toxic action on proteins and other cytoplasmic constituents. Citrate could then subsequently move into the vacuole. Chelation of zinc may be important in its movement across the tonoplast. These speculations are further added to when it is considered how zinc stimulates citrate synthesis in tolerant plants. The stimulation of citrate synthesis by zinc is clearly different from that shown by nitrate which often stimulates organic acid synthesis in plants. Rankin and Thurman (unpublished data) working with *D. caespitosa* and attempting to convert a non-tolerant clone into a tolerant one by growth in high nitrate (12 mM) media, have shown that this treatment stimulates the synthesis of several commonly occurring acids. In contrast, when the zinc-tolerant clone is grown in the presence of zinc, only citrate and, to a much lesser extent, malate increase, suggesting that different mechanisms are involved.

If acids such as citrate and malate are involved in tolerance, it is unlikely that these molecules themselves are responsible for the specificity of tolerance. If the stability constants of citrate chelates are compared, it would be expected that a plant showing copper tolerance would also show zinc tolerance, but it is known that this is not always the case. It is not difficult to suggest that the specificity may reside somewhere in the enzymic machinery used in the production of malate or citrate.

6. ENZYMIC ADAPTATIONS

A number of plant species, including edaphic ecotypes, have been shown to possess enzymes which are not inhibited by heavy metals. For higher plants, it is still not possible to say whether or not such modifications play a direct role in tolerance, or whether they are involved in some other aspect of the physiology of the plant which is related to certain factors in the habitat apart from the presence of heavy metals.

Woolhouse (1970) initiated research on acid phosphatases in edaphic ecotypes. In a study of the inhibition of root growth by a range of different metals, he argued that cell elongation depends partly upon synthesis of new cell wall materials and suggested that the *p*-nitrophenylphosphatases may be involved in polysaccharide synthesis. In the case of root cells, these enzymes will come under the influence of any metal present in the soil solution. Wainwright and Woolhouse (1975) studied the inhibitions by copper of cell wall acid phosphatases of a copper-tolerant clone and a non-tolerant clone of *A. tenuis*. By measurement of inhibitor constants they were able to show that copper, a competitive inhibitor in this case, formed stronger complexes with the enzymes of the non-tolerant clone than the tolerant clone. The levels of copper required to give significant inhibitions were of the same order of magnitude as the level of copper in the soil solution (approximately 0·025 mM) of the plant's habitat. This was interpreted as a modification with adaptive significance in the copper-rich habitat. In a similar study for zinc, these workers could not conclude that inhibition of cell wall acid phosphatases by zinc constituted a selective pressure. However, this may not be so for the soluble acid phosphatases. Wainwright and Woolhouse (1978), utilizing a previous observation of Cox and Thurman (1978) on the effect of zinc upon these enzymes, suggested that modification of these soluble acid phosphatases may have been a pre-requisite of evolution of zinc tolerance. All of the studies reported so far were made using *in vitro* methods. Mathys (1975) has made a comparative study using *in vivo* and *in vitro* methods on the effects of a number of metals on various enzymic activities present in tolerant and non-tolerant clones of *Silene cucubalis*. Using *in vitro* methods, he could not distinguish any differences in the degree of inhibition shown by non-tolerant and tolerant plants. Using *in vivo* methods, differences were detected which could be attributed to a cellular compartmentalization of metal in vacuoles of the type already discussed, and not due to differences in enzymic tolerances.

Genetical studies have a bearing on the possible number of enzymes which may be modified in metal tolerant plants. MacNair (1977), from data obtained with copper-tolerant *Mimulus guttatus*, has evidence for the involvement of only two major genes in tolerance, though he pointed out that other loci of lesser effect may also be involved. MacNair argued that if enzymic adaptation formed a major part of the tolerance mechanism, many genes would be involved because heavy metals inhibit the activity of many enzymes. He further suggested that, unless one enzyme system is much more susceptible to copper than others, the substitution of a single non-tolerant enzyme by a tolerant one is unlikely to cause a significant increase in tolerance. A much greater contribution to tolerance would be made by the production of a substance(s) which was able to bind metals and so reduce their cytoplasmic toxicity. Even if tolerant plants possess such a primary mechanism, their cytoplasm may still, at times, contain elevated levels of metal, and as MacNair suggested, it could be that selection will act to increase the tolerance of a particular enzyme which may account for the involvement of other loci of minor importance and give rise to a secondary tolerance mechanism.

7. PERMEABILITY EFFECTS

Heavy metals cause leakage of a number of materials from plants, an effect often attributed to an increase in membrane permeability. Most studies of this type have been concerned with nett leakage, usually of potassium, from root cells. The leakage has not been resolved into any possible differential effects of zinc upon the influx or efflux of potassium. Wyn Jones and Sutcliffe (1972) and Brookes and Collins (unpublished data) have all observed a greater effect of zinc upon the leakage of potassium from non-tolerant rather than tolerant roots of grasses, though Wainwright and Woolhouse (1977) using a zinc-tolerant clone of *A. tenuis* found that zinc was without effect upon potassium release from the roots of either clone. In contrast, all concentrations of copper tested caused some release of potassium, but this effect was less with the copper-tolerant clone than with either the zinc-tolerant or non-tolerant clones. Wainwright and Woolhouse (1977) suggested that potassium leakage may be only an incidental event in the inhibition of root growth by copper, rather than a crucial part of the mechanism of inhibition.

Relationships between calcium and the toxic effects of heavy metals have been investigated by a number of workers (Baker, 1978*b*), though how these relate to possible tolerance mechanisms is not clear. Calcium has an ameliorating effect on the inhibition of root growth and also on potassium leakage from roots caused by heavy metal treatment (Wyn Jones and Lunt, 1967), though this is not always so. In a non-tolerant clone of *A. tenuis* Wyn Jones and Sutcliffe (1972) found enhanced leakage of potassium in the presence of zinc and calcium rather than with zinc alone.

The preceding discussion illustrates the complex nature of the interactions between metal-tolerant plants and soil-borne metals. At present, no precise answer to the question of mechanisms of tolerance can be advanced; indeed, in the case of certain elements (e.g. lead), very little relevant information is available. Further progress in elucidating the underlying mechanism behind this unique biological phenomenon will depend on using an experimental plant species suitable not only for biochemical and physiological investigations, but which also possesses the potential for further genetic studies.

REFERENCES

Baker, A. J. M. (1978*a*). Ecophysiological aspects of zinc tolerance in *Silene maritima* With. *New Phytol.* **80**: 635–42.

Baker, A. J. M. (1978*b*). The uptake of zinc and calcium from solution culture by zinc-tolerant and non-tolerant *Silene maritima* With. in relation to calcium supply. *New Phytol.* **81**: 321–30.

Brookes, A. (1979). Ion and water relations of zinc tolerant grass clones. Ph.D. Thesis, University of Liverpool, UK.

Brookes, A., J. C. Collins and D. A. Thurman (1980). The mechanism of zinc tolerance in grasses. *J. Plant Nutr.* (in press).

Cox, R. M. and D. A. Thurman (1978). Inhibition by zinc of soluble and cell wall acid phosphatases of zinc-tolerant and non-tolerant clones of *Anthoxanthum odoratum*. *New Phytol.* **80**: 17–22.

Ernst, W. (1976). Physiological and biochemical aspects of metal tolerance. In: *Effects of Air Pollutants on Plants* (T. A. Mansfield (ed)), Cambridge University Press, Cambridge, pp. 115–33.

Grime, J. P. and J. G. Hodgson (1969). An investigation of the significance of lime chlorosis by means of large scale comparative experiments. In: *Ecological Aspects of the Mineral Nutrition of Plants*, 9th Symposium, British Ecological Society (I. H. Rorison, A. D. Bradshaw and M. J. Chadwick (eds)), Blackwell, Oxford, pp. 67–99.

Gund, F. R. N. and P. E. Wilcox (1956). Complex formation between metallic cations and proteins, peptides and amino acids. *Adv. Protein Chem.* **11**: 311–427.

Jones, L. H. (1961). Aluminium uptake and toxicity in plants. *Plant Soil.* **13**: 297–310.

Lee, J., R. D. Reeves, R. R. Brooks and T. Jaffre (1978). The relation between nickel and citric acid in some nickel-accumulating plants. *Phytochem.* **17**: 1033–5.

MacNair, M. R. (1977). Major genes for copper tolerance in *Mimulus guttatus*. *Nature, London.* **268**: 428–30.

Mathys, W. (1975). Enzymes of heavy-metal-resistant and non-resistant populations of *Silene cucubalus* and their interactions with some heavy metals *in vitro* and *in vivo*. *Physiol. Plant.* **33**: 161–5.

Mathys, W. (1977). The role of malate, oxalate and mustard oil glucosides in the evolution of zinc-resistance in herbage plants. *Physiol. Plant.* **40**: 130–6.

Peterson, P. J. (1969). The distribution of Zinc-65 in *Agrostis tenuis* Sibth. and *A. stolonifera* L. tissues. *J. Exp. Bot.* **20**: 863–74.

Robards, A. W., S. M. Jackson, D. T. Clarkson and J. Sanderson (1973). The structure of barley roots in relation to the transport of ions into the stele. *Protoplasma.* **77**: 291–311.

Thomas, M., S. L. Ranson and J. A. Richardson (1973). *Plant Physiology.* 5th Edition, Longman, London. 451 pp.

Tiffin, L. O. (1972). Translocation of micronutrients in plants. In *Micronutrients in Agriculture* (J. J. Mortvedt, P. M. Giordano and W. L. Lindsay (eds)), Soil Science Society America, Madison, Wisconsin, pp. 389–418.

Turner, R. G. (1967). Experimental studies on heavy metal tolerance. Ph.D. Thesis, University of Wales, Cardiff, UK.

Turner, R. G. (1969). Heavy metal tolerance in plants. In: *Ecological Aspects of the Mineral Nutrition of Plants*, 9th Symposium, British Ecological Society (I. H. Rorison, A. D. Bradshaw and M. J. Chadwick (eds)), Blackwell, Oxford, pp. 399–410.

Turner, R. G. and C. Marshall (1971). The accumulation of ^{65}Zn by root homogenates of zinc-tolerant and non-tolerant clones of *Agrostis tenuis* Sibth. *New Phytol.* **70**: 539–45.

Turner, R. G. and C. Marshall (1972). The accumulation of zinc by subcellular fractions of roots of *Agrostis tenuis* Sibth. in relation to zinc tolerance. *New Phytol.* **71**: 671–6.

Wainwright, S. J. and H. W. Woolhouse (1975). Physiological mechanisms of heavy metal tolerance in plants. In: *The Ecology of Resource Degradation and Renewal*, 15th Symposium, British Ecological Society (M. J. Chadwick and G. T. Goodman (eds)), Blackwell, Oxford, pp. 231–59.

Wainwright, S. J. and H. W. Woolhouse (1977). Some physiological aspects of copper and zinc tolerance in *Agrostis tenuis* Sibth.: Cell elongation and membrane damage. *J. Exp. Bot.* **28**: 1029–36.

Wainwright, S. J. and H. W. Woolhouse (1978). Inhibition by zinc of cell wall acid phosphatases from roots in zinc-tolerant and non-tolerant clones of *Agrostis tenuis* Sibth. *J. Exp. Bot.* **29**: 525–31.

West, P. W. (1979). Release to shoots of copper stored in roots of *Pinus radiata* D. Don. *Annal. Bot.* **43**: 237–40.

Woolhouse, H. W. (1970). Environment and enzyme evolution in plants. In: *Phytochemical Phylogeny* (J. B. Harborne (ed)), Academic Press, London and New York, pp. 207–29.

Wu, L. and J. Antonovics (1975). Zinc and copper uptake by *Agrostis stolonifera*, tolerant to both zinc and copper. *New Phytol.* **75**: 231–7.

Wu, L., D. A. Thurman and A. D. Bradshaw (1975). The uptake of copper and its effect upon respiratory processes of roots of copper-tolerant and non-tolerant clones of *Agrostis stolonifera*. *New Phytol.* **75**, 225–9.

Wyn Jones, R. G. and O. R. Lunt (1967). The function of calcium in plants. *Bot. Rev.* **33**: 407–26.

Wyn Jones, R. G. and M. H. Sutcliffe (1972). Some physiological aspects of heavy metal tolerance in *Agrostis tenuis*. Welsh Soils Discussion Group Report, pp. 1–15.

Wyn Jones, R. G., M. Sutcliffe and C. Marshall (1971). Physiological and biochemical basis for heavy metal tolerance in clones of *Agrostis tenuis*. In: *Recent Advances in Plant Nutrition, Vol. 2* (R. M. Samish (ed)), Gordon and Breach, New York, pp. 575–81.

Index